Cover photo: Courtesy of Dupont

Current Intelligence Bulletin 66

Derivation of Immediately Dangerous to Life or Health (IDLH) Values

DEPARTMENT OF HEALTH AND HUMAN SERVICES
Centers for Disease Control and Prevention
National Institute for Occupational Safety and Health

> This document is in the public domain and may be freely copied or reprinted.

Disclaimer

Mention of any company or product does not constitute endorsement by the National Institute for Occupational Safety and Health (NIOSH). In addition, citations to websites external to NIOSH do not constitute NIOSH endorsement of the sponsoring organizations or their programs or products. Furthermore, NIOSH is not responsible for the content of these websites. All Web addresses referenced in this document were accessible as of the publication date.

Ordering Information

To receive documents or other information about occupational safety and health topics, contact NIOSH:

   Telephone: 1-800-CDC-INFO (1-800-232-4636)
   TTY: 1-888-232-6348
   CDC INFO: www.cdc.gov/info

or visit the NIOSH website at www.cdc.gov/niosh.

For a monthly update on news at NIOSH, subscribe to *NIOSH eNews* by visiting www.cdc.gov/niosh/eNews.

Suggested Citation

NIOSH [2013]. Current intelligence bulletin 66: derivation of immediately dangerous to life or health (IDLH) values. Cincinnati, OH: US Department of Health and Human Services, Centers for Disease Control and Prevention, National Institute for Occupational Safety and Health, DHHS (NIOSH) Publication 2014-100.

DHHS (NIOSH) Publication No. 2014-100

November 2013

SAFER • HEALTHIER • PEOPLE™

Foreword

Since the establishment of the original Immediately Dangerous to Life or Health (IDLH) values in 1974, the National Institute for Occupational Safety and Health (NIOSH) has continued to review available scientific data to improve the methodology used to derive acute exposure guidelines, in addition to the chemical-specific IDLH values. The primary objective of this Current Intelligence Bulletin (CIB) is to present a methodology, based on the modern principles of risk assessment and toxicology, for the derivation of IDLH values, which characterize the health risks of occupational exposures to high concentrations of airborne contaminants. The methodology for deriving IDLH values presented in the CIB incorporates the approach established by the National Advisory Committee on Acute Exposure Guideline Levels (AEGLs) for Hazardous Substances—consisting of members from the U.S. Environmental Protection Agency, U.S. Department of Defense, U.S. Department of Energy, U.S. Department of Transportation, other federal and state government agencies, the chemical industry, academia, labor, and other organizations from the private sector—during the derivation of community-based acute exposure limits. The inclusion of the AEGL methodology has helped ensure that the IDLH values derived with use of the guidance provided in this document are based on validated scientific rationale.

The intent of this document is not only to update the IDLH methodology used by NIOSH to develop IDLH values based on contemporary risk assessment practices, but also to increase the transparency behind their derivation. The increased transparency will provide occupational health professionals, risk managers, and emergency response personnel additional information that can be applied to improve characterization of the hazards of high concentrations of airborne contaminants. This will also facilitate a more informed decision-making process for the selection of respirators and establishment of risk management plans for non-routine work practices and emergency preparedness plans capable of better protecting workers.

John Howard, M.D.
Director, National Institute for Occupational
  Safety and Health
Centers for Disease Control and Prevention

This page intentionally left blank.

Executive Summary

Chemicals are a ubiquitous component of the modern workplace. Occupational exposures to chemicals have long been recognized as having the potential to adversely affect the lives and health of workers. Acute or short-term exposures to high concentrations of some airborne chemicals have the ability to quickly overwhelm workers, resulting in a spectrum of undesirable outcomes that may include irritation of the eyes and respiratory tract, severe irreversible health effects, impairment of the ability to escape from the exposure environment, and, in extreme cases, death. Airborne concentrations of chemicals capable of causing such adverse health effects or of impeding escape from high-risk conditions may arise from a variety of non-routine workplace situations affecting workers, including special work procedures (e.g., in confined spaces), industrial accidents (e.g., chemical spills or explosions), and chemical releases into the community (e.g., during transportation incidents or other uncontrolled-release scenarios).

Since the 1970s, the National Institute for Occupational Safety and Health (NIOSH) has been responsible for the development of acute exposure guidelines called Immediately Dangerous to Life or Health (IDLH) values, which are intended to characterize these high-risk conditions. Used initially as key components of the *NIOSH Respirator Selection Logic* [NIOSH 2004], IDLH values are established (1) to ensure that the worker can escape from a given contaminated environment in the event of failure of the respiratory protection equipment and (2) to indicate a maximum level above which only a highly reliable breathing apparatus, providing maximum worker protection, is permitted. In addition, occupational health professionals have employed these acute exposure guidelines beyond their initial purpose as a component of the *NIOSH Respirator Selection Logic*. Examples of such applications of the IDLH values include the development of Risk Management Plans (RMPs) for non-routine work practices governing operations in high-risk environments (e.g., confined spaces) and the development of Emergency Preparedness Plans (EPPs), which provide guidance for emergency response personnel and workers during unplanned exposure events.

Since the establishment of the IDLH values in the 1970s, NIOSH has continued to review available scientific data to improve the protocol used to derive acute exposure guidelines, in addition to the chemical-specific IDLH values. The information presented in this Current Intelligence Bulletin (CIB) represents the most recent update of the scientific rationale and the methodology (hereby referred to as the IDLH methodology) used to derive IDLH values. The primary objectives of this document are to

- Provide a brief history of the development of IDLH values
- Update the scientific bases and risk assessment methodology used to derive IDLH values from quality data
- Provide transparency behind the rationale and derivation process for IDLH values
- Demonstrate how scientifically credible IDLH values can be derived from available data resources.

The IDLH methodology outlined in this CIB reflects the modern principles and understanding in the fields of risk assessment, toxicology, and occupational health and provides the scientific rationale for the derivation of IDLH values based on contemporary risk assessment practices. According to this protocol, IDLH values are based on health effects considerations determined through a critical assessment of the toxicology and human health effects data. This approach ensures that the IDLH values reflect an airborne concentration of a substance that represents a high-risk situation that may endanger workers' lives or health. Relevant airborne concentrations are typically addressed through the characterization of inhalation exposures; however, airborne chemicals can also contribute to toxicity through other exposure routes, such as the skin and eyes. In this document, airborne concentrations are referred to as *acute inhalation limits or guidelines* to adhere to commonly used nomenclature.

The emphasis on health effects is consistent with both the traditional use of IDLH values as a component of the respirator selection logic and the growing applications of IDLH values in RMPs for non-routine work practices governing operations in high-risk environments (e.g., confined spaces) and the development of EPPs. Incorporated in the IDLH methodology are the standing guidelines and procedures [NAS 2001] used for the development of community-based acute exposure limits called Acute Exposure Guideline Levels (AEGLs). The inclusion of the AEGL methodology has helped ensure that the health-based IDLH values derived with use of the guidance provided in this document are based on validated scientific rationale.

The IDLH methodology is based on a weight-of-evidence approach that applies scientific judgment for critical evaluation of the quality and consistency of scientific data and in extrapolation from the available data to the IDLH value. The weight-of-evidence approach refers to critical examination of all available data from diverse lines of evidence and the derivation of a scientific interpretation on the basis of the collective body of data, including its relevance, quality, and reported results. This is in contrast to a purely hierarchical or strength-of-evidence approach, which relies on rigid decision criteria for selecting a critical adverse effect, a point of departure (POD), or the point on the dose–response curve from which dose extrapolation is initiated and for applying default uncertainty factors (UFs) to derive the IDLH value. Conceptually, the derivation process for IDLH values is similar to that used in other risk-assessment applications, including these steps:

- Hazard characterization
- Identification of critical adverse effects
- Identification of a POD
- Application of appropriate UFs, based on the study and POD
- Determination of the final risk value.

However, the use of a weight-of-evidence approach allows for integration of all available data that may originate from different lines of evidence into the analysis and the subsequent derivation of an IDLH value. Ideally, this ensures that the analysis is not restricted to a limited dataset or a single study for a specific chemical. In particular, application of the appropriate UFs to each potential POD allows for consideration of the impact of the overall dataset as well as the uncertainties associated with each potential key study in determining the final IDLH value.

The primary steps (*see Figure 3–1*) applied in the establishment of an IDLH value include the following:

- Critical review of human and animal toxicity data to identify potential relevant studies and characterize the various lines of evidence that can support the derivation of the IDLH value

- Determination of a chemical's mode of action (MOA) or description of how a chemical exerts its toxic effects

- Application of duration adjustments (time scaling) to determine 30-minute-equivalent exposure concentrations and the conduct of other dosimetry adjustments, as needed

- Selection and application of a UF for POD or critical adverse effect concentration, identified from the available studies to account for issues associated with interspecies and intraspecies differences, severity of the observed effects, data quality, or data insufficiencies

- Development of the final recommendation for the IDLH value from the various alternative lines of evidence, with use of a weight-of-evidence approach to all of the data.

NIOSH recognizes that in some cases a health-based IDLH value might not account for all workplace hazards, such as safety concerns or considerations. Here are some examples of situations and conditions that might preclude the use of a health-based IDLH value:

- The airborne concentration of a substance is sufficient to cause oxygen deprivation (oxygen concentration <19.5%), a life-threatening condition

- The concentration of particulate matter generated during a process significantly reduces visibility, preventing escape from the hazardous environment

- The airborne concentration of a gas or vapor is greater than 10% of the lower explosive limit (LEL) and represents an explosive hazard.

In such cases, it is important that safety hazards or other considerations be taken into account. Information on the safety hazards will be incorporated in the support documentation (*see Appendix A*) for an IDLH value, to aid occupational health professionals in the development of RMPs for non-routine work practices governing operations in high-risk environments (e.g., confined spaces) and EPPs. In the event that the derived health-based IDLH value exceeds 10% of the LEL concentration for a flammable gas or vapor, the air concentration that is equal to 10% of the LEL will become the default IDLH value for the chemical. The following hazard statement will be included in the support documentation: "The health-based IDLH value is greater than 10% of the LEL (>10% LEL) of the chemical of interest in the air. Safety considerations related to the potential hazard of explosion must be taken into account." In addition, the notation ">10% LEL" will appear beside the IDLH value in the *NIOSH Pocket Guide to Chemical Hazards* [NIOSH 2005] and other NIOSH publications. The equivalent default approach for dust would be based on 10% of the minimum explosive concentration (MEC). However, determining the combustibility of dusts is complicated and dictated by the relationship between multiple dust-specific factors including, but not limited to, particle size distribution, minimum ignition energy, explosion intensity, and dispersal in the air [Cashdollar 2000]. The ability to quantify dust-specific concentrations that could represent explosive hazards for risk assessment purposes is limited and often not possible given the absence of critical data, such as chemical-specific MEC and other previously identified factors. Despite the absence of specific guidance, NIOSH will critically assess the explosive nature of a dust when sufficient technical data are available. If determined to be appropriate, the findings of this assessment will be incorporated into the derivation process to ensure that the IDLH value is protective against both health and safety hazards. When an explosive hazard is identified for an aerosol, NIOSH will include the following hazard statement: "Dust may represent an explosive hazard. Safety considerations related to hazard of explosion must be taken into account." In addition, the notation (Combustible Dust) will appear in other NIOSH publications.

Supplemental information is included in this CIB to provide insight into (1) the literature search strategy, (2) the scheme used to prioritize and select chemicals for which an IDLH value will be established, and (3) an overview of the analysis applied by NIOSH to develop a scientifically based approach for the selection of the UF during the derivation of IDLH values. In addition, Appendix A presents an example of the derivation of an IDLH value for chlorine (CAS# 7782-50-5), based on the scientific rationale and process outlined in this CIB. The example highlights the primary steps in establishment of an IDLH value, including a critical review of the identified human and animal data, discussion of the selection of the POD and UF, and extrapolation of the 30-minute-equivalent exposure concentration from animal toxicity data.

Contents

Foreword	iii
Executive Summary	v
Abbreviations and Acronyms	xi
Glossary	xv
Acknowledgements	xxiii
1 Introduction	1
1.1 Background	1
1.2 The Standards Completion Program	2
1.3 Basis of the Original IDLH Values	3
1.4 Update of the IDLH Values in 1994	4
1.5 Purpose and Objectives of the IDLH Values	5
2 Comparison of IDLH Values to Alternative Short-term Exposure Limits/Values	7
2.1 Acute Exposure Guideline Levels	7
2.2 Emergency Response Planning Guidelines	9
2.3 Occupational Exposure Limits	10
2.4 Other Acute Exposure Limits/Values	12
3 Criteria for Determining IDLH Values	13
3.1 Importance of Mode of Action and Weight-of-Evidence Approach	15
3.2 Process for Prioritization of Chemicals	17
3.3 Literature Search Strategy	17
3.4 Determining the Critical Study and Endpoint	17
3.4.1 Study Quality Considerations	19
3.4.2 Study Relevance Considerations	19
3.5 Time Scaling	31
3.6 Inclusion of Safety Considerations	33
4 Use of Uncertainty Factors	35
4.1 Application of Uncertainty Factors	35
4.2 The NIOSH IDLH Value Uncertainty Factor Approach	36
4.3 Research Support for the NIOSH Uncertainty Factor Approach	38
References	41
Appendix A: Example of the Derivation of an IDLH Value	45
A.1 Support Documentation for the Revised IDLH Value for Chlorine	45
A.2 Animal Toxicity Data	45
A.3 Human data	47
A.4 IDLH Value Rationale Summary	48
References	48
Appendix B: IDLH Value Development Prioritization	51

    References 54

Appendix C: Critical Effect Determination for IDLH Value Development—
  Consideration of Severity, Reversibility, and Impact on Escape Impairment 55
    References 65

Appendix D: Analyses Supporting the Development of Uncertainty
  Factor Approach 67
    D.1 Analysis for Selected Approach............................ 67
    D.2 Recommendation for Deriving IDLH Values 69
    References 71

Appendix E: Quantitative Adjustments during the Derivation of IDLH Values 73
    E.1 Inhalation Volume Adjustments Approach for
           Route-to-Route Extrapolation 73
    E.2 Time Scaling Adjustments 75
    References 82

Abbreviations and Acronyms

ACGIH	American Conference of Governmental Industrial Hygienists
AIHA	American Industrial Hygiene Association
AEGL	Acute Exposure Guideline Level (published by NRC)
AUC	area under the curve
ATSDR	Agency for Toxic Substances and Disease Registry
BBDR	biologically based dose response
BMC	benchmark concentration
BMCL	benchmark concentration lower-bound confidence limit
BMD	benchmark dose
BMDS	Benchmark Dose Software (developed by USEPA)
"C"	ceiling value
CA	carcinogen
Cal/EPA	California Environmental Protection Agency
CAS#	Chemical Abstracts Service Registry Number
CDC	Centers for Disease Control and Prevention
CFATS	Chemical Facility Anti-Terrorism Standards (developed by DHS)
CFR	Code of Federal Regulations
CHEMID	online chemical identification database (developed by NLM)
CIB	Current Intelligence Bulletin (developed by NIOSH)
C_{max}	peak (maximum) concentration
CNS	central nervous system
COHb	carboxyhemoglobin
Conc	concentration
DHHS	U.S. Department of Health and Human Services
DHS	U.S. Department of Homeland Security
DOD	U.S. Department of Defense
DOE	U.S. Department of Energy
DOL	U.S. Department of Labor
DOT	U.S. Department of Transportation
DT	developmental toxicant
EC	effective concentration

EEGL	Emergency and Continuous Exposure Guidance Levels (published by NRC)
EINECS	European INventory of Existing Commercial chemical Substances
EMBASE	online biomedical journal abstract and indexing database (subscription based)
EPP	Emergency Preparedness Plan
ERPG	Emergency Response Planning Guidelines (developed by AIHA)
ERG	Emergency Response Guidebook (developed by DOT)
EU	European Union
FEL	frank effect level
FACA	Federal Advisory Committee Act
GI	gastrointestinal
GLP	Good Laboratory Practices
HAZARDTEXT®	online hazardous substance database (subscription based)
HazMap	online occupational exposure to hazardous agents database (developed by NLM)
HCN	hydrogen cyanide
hr	hour
HPV	high production volume
HSDB	Hazardous Substance Data Bank (developed by NLM)
HSEES	Hazardous Substance Emergency Events Surveillance (developed by ATSDR)
IARC	International Agency for Research on Cancer
ICSC	International Chemical Safety Cards (developed by IPCS)
IDLH	Immediately Dangerous to Life or Health (developed by NIOSH)
i.p.	intraperitoneal
IPCS	International Programme on Chemical Safety
IRIS	Integrated Risk Information System (developed by USEPA)
IRR	irritant
ITER	International Toxicity Estimates for Risk database (developed by TERA)
JSC	Johnson Space Center (division of NASA)
k	a constant reflected in equations expressing "conc × time" relationships
kg	kilogram
L	liter
LC	lethal concentration

LD	lethal dose
LEL	lower explosive limit
L/min	liters per minute
LOAEL	lowest observed adverse effect level
LOEL	lowest observed effect level
m^3	cubic meter
MEC	minimum explosive concentration
MEDITEXT®	online medical and toxicology database (subscription based)
mg/m^3	milligrams per cubic meter of air
mg/m^3-min	milligrams per cubic meter of air per minute
min	minute
MOA	mode of action
MSHA	Mine Safety and Health Administration
NAC/AEGL	National Advisory Committee for Acute Exposure Guideline Levels for Hazardous Substances
NASA	National Aeronautics and Space Administration
NAS/NRC	National Academy of Sciences/National Research Council
NIOSH	National Institute for Occupational Safety and Health
NIOSHTIC2	bibliographic database of NIOSH-supported occupational safety and health publications
NJ-HSFS	New Jersey Hazardous Substance Fact Sheets
NLM	National Library of Medicine
NOAEL	no observed adverse effect level
NOEL	no observed effect level
NRC	National Research Council
NTP	National Toxicology Program
OECD	Organisation for Economic Co-operation and Development
OEL	occupational exposure limit
OSHA	Occupational Safety and Health Administration
PAL	Provisional Advisory Levels (developed by DHS)
PBPK	physiologically based pharmacokinetic
PEL	Permissible Exposure Limit (developed by OSHA and MSHA)
ppm	parts per million
POD	point of departure
PUBMED	online biomedical literature citation database (developed by NLM)

RD	respiratory depression
REL	Recommended Exposure Limit (developed by NIOSH)
RfC	inhalation reference concentration
RIVM	Netherlands National Institute for Public Health and the Environment
RMP	Risk Management Plan
R-phrases	risk phrases (developed by EU)
RTECS	Registry of Toxic Effects of Chemical Substances
SCAPA	Subcommittee on Consequence Assessment and Protective Actions
SCBA	self-contained breathing apparatus
SCP	Standards Completion Program (developed by NIOSH and OSHA)
SMAC	Spacecraft Maximum Allowable Concentration (developed by NASA, published by NRC)
SOP	Standing Operating Procedures
SPEGL	Short-term Public Emergency Guidance Levels (developed by NRC)
STEG	short-term exposure guidelines
STEL	Short Term Exposure Limit
ST	short-term exposure limit
TEEL	Temporary Emergency Exposure Limit (developed by DOE)
TERA	Toxicology Excellence for Risk Assessment
TIH	toxic inhalation hazard (developed by DOT)
TLV®	Threshold Limit Value (developed by ACGIH)
TOXLINE	online toxicology literature database (developed by NLM)
TWA	time-weighted average
UF	uncertainty factor
USEPA	U.S. Environmental Protection Agency
WEEL	Workplace Environmental Exposure Limits (developed by AIHA)
WHO	World Health Organization

Glossary*

Acute Exposure: Exposure by the oral, dermal, or inhalation route for 24 hours or less.

Acute Exposure Guideline Levels (AEGLs): Threshold exposure limits for the general public applicable to emergency exposure periods ranging from 10 minutes to 8 hours. AEGL-1, AEGL 2, and AEGL-3 are developed for five exposure periods (10 and 30 minutes, 1 hour, 4 hours, and 8 hours) and are distinguished by varying degrees of severity of toxic effects ranging from transient, reversible effects to life-threatening effects [NAS 2001]. AEGLs are intended to be guideline levels used during rare events or single once-in-a-lifetime exposures to airborne concentrations of acutely toxic, high-priority chemicals [NAS 2001]. The threshold exposure limits are designed to protect the general population, including the elderly, children or other potentially sensitive groups that are generally not considered in the development of workplace exposure recommendations (additional information available at http://www.epa.gov/oppt/aegl/).

Acute Reference Concentration (RfC): An estimate (with uncertainty spanning perhaps an order of magnitude) of a continuous inhalation exposure for an acute duration (24 hours or less) of the human population (including sensitive subgroups) that is likely to be without an appreciable risk of deleterious effects during a lifetime. It can be derived from a NOAEL, LOAEL, or benchmark concentration, with uncertainty factors (UFs) generally applied to reflect limitations of the data used. Generally used in USEPA noncancer health assessments [USEPA 2010].

Acute Toxicity: Any poisonous effect produced within a short period of time following an exposure, usually 24 to 96 hours.

Acute Toxicity Test: Experimental animal study to determine what adverse effects occur in a short time (usually up to 14 days) after a single dose of a chemical or after multiple doses given in up to 24 hours.

Adverse Effect: A substance-related biochemical change, functional impairment, or pathologic lesion that affects the performance of an organ or system or alters the ability to respond to additional environmental challenges.

Analytical (Actual) Concentration: The test article concentration to which animals are exposed (i.e., the concentration in the animals' breathing zone), as measured by analytical (GC, HPLC, etc.) or gravimetric methods. The analytical or gravimetric concentration (not the nominal concentration) is usually used for concentration response assessment.

Assigned Protection Factor (APF): The minimum anticipated protection provided by a properly functioning respirator or class of respirators to a given percentage of properly

---

*Except where specific references are given, glossary definitions are from numerous sources such as AIHA [2008], Hayes [2008], IUPAC [2007], NAS [1986, 2001], NASA [1999], NIOSH [2005], OSHA [2003], US DHS [2007], US DOE [2008], and US DOT [2008].

fitted and trained users. For example, an APF of 10 for a respirator means that a user could expect to inhale no more than one tenth of the airborne contaminant present.

Benchmark Dose/Concentration (BMD/BMC): A dose or concentration that produces a predetermined change in response rate of an effect (called the benchmark response, or BMR) compared to background [USEPA 2010] (additional information available at http://www.epa.gov/ncea/bmds/).

Benchmark Response (BMR): A predetermined change in response rate of an effect. Common defaults for the BMR are 10% or 5%, reflecting study design, data variability, and sensitivity limits used.

BMCL: A statistical lower confidence limit on the concentration at the BMC [USEPA 2010].

Biologically Based Dose Response (BBDR) model: A predictive model that describes biological processes at the cellular and molecular level, linking the target organ dose to the adverse effect [USEPA 2010].

Bolus Exposure: A single, relatively large dose.

Bounding: A process of identifying estimates of exposure, dose, or risk that are clearly higher than or lower than the exposure, dose, or risk of interest. Bounding can help to define the practical uncertainty associated with the estimate of a derived risk value, such as an IDLH value.

Cancer Risk: The likelihood of developing cancer, given a specific exposure (i.e., during a working lifetime). Individual cancer risks are determined by multiplying a specific exposure by the cancer potency. A 10^{-3} risk level is often characterized as a 1 in 1,000 chance of developing cancer in occupational risk assessment.

Carcinogen: An agent capable of causing cancer.

Carcinogenicity: Process of induction of malignant tumors by chemical, physical, or biological agents.

Ceiling Value ("C"): U.S. term in occupational exposure indicating the airborne concentration of a potentially toxic substance that should never be exceeded in a worker's breathing zone.

Chronic Exposure: Repeated exposure for an extended period of time. Typically exposures are more than approximately 10% of life span for humans and >90 days to 2 years for laboratory species.

Concentration (Conc): The mass of test article per unit volume of air (e.g., mg/L, mg/m^3) or the volume of test article per unit volume (e.g., ppm, mL/L).

Concentration-response Curve: Graph of the relationship between the exposure concentration and the incidence or other measure of response of a defined biological effect in an exposed population or animal study.

Critical Study: The study that contributes most significantly to the qualitative and quantitative assessment of risk [USEPA 2010].

Cumulative Toxicity: Toxicity that is related to the cumulative, or total, dose to an organ or the body of an individual, up to a specified date or time.

Developmental Toxicity: Adverse effects on the developing organism that may result from exposure prior to conception (either parent), during prenatal development, or postnatally until the time of sexual maturation [USEPA 2010]. The major manifestations of developmental toxicity include death of the developing organism, structural abnormality, altered growth, and functional deficiency.

De Novo: Referring to an analysis that does not build on prior analyses.

Dose: The amount of a substance available for interactions with metabolic processes or biologically significant receptors after crossing the outer boundary of an organism [USEPA 2010].

Dosimetry: Estimating or measuring the quantity of material at specific target sites; determination of respiratory tract region deposition fractions.

ECt_{50}: A combination of the effective concentration of a substance in the air and the exposure duration that is predicted to cause an effect in 50% (one half) of the experimental test subjects.

Emergency and Continuous Exposure Guidance Level (EEGL): A ceiling guidance level for unpredicted, single, short-term, emergency exposures (1 to 24 hours) of a defined occupational group. EEGLs are developed at the request of the U.S. Department of Defense by the National Research Council's Committee on Toxicology [NAS 1996, 2008].

Emergency Response Planning Guidelines (ERPGs): Maximum airborne concentrations below which nearly all individuals can be exposed without experiencing health effects for 1-hour exposure. ERPGs are presented in a tiered fashion with health effects ranging from mild or transient to serious, irreversible, or life threatening (depending on the tier). ERPGs are developed by the American Industrial Hygiene Association [AIHA 2006].

Endpoint: An observable or measurable biological event or sign of toxicity ranging from biomarkers of initial response to gross manifestations of clinical toxicity.

Exposure: Contact made between a chemical, physical, or biological agent and the outer boundary of an organism. Exposure is quantified as the amount of an agent available at the exchange boundaries of the organism (e.g., skin, lungs, gut).

Extrapolation: An estimate of the response at a point outside the range of the experimental data, generally through the use of a mathematical model, although qualitative extrapolation may also be conducted. The model may then be used to extrapolate to response levels that cannot be directly observed.

Fetal Toxicity: An adverse effect occurring in the fetus from exposure to a substance. These effects can occur through direct interaction with the fetus or indirectly from the effects of maternal toxicity.

Gestation: Pregnancy, the period of development in the uterus from conception until birth.

Hazard: A potential source of harm. Hazard is distinguished from risk, which is the probability of harm under specific exposure conditions.

Healthy Worker Effect: Epidemiological phenomenon observed initially in studies of occupational diseases: workers usually exhibit lower overall disease and death rates than the general population, due to the fact that elderly individuals and those with significant pre-existing illness are less likely to be active in the workforce than healthy adults. Death rates in the general population may be inappropriate for comparison with occupational death rates, if this effect is not taken into account.

Immediately Dangerous to Life or Health (IDLH) condition: A situation that poses a threat of exposure to airborne contaminants when that exposure is likely to cause death or immediate or delayed permanent adverse health effects or prevent escape from such an environment [NIOSH 2004].

IDLH value: A maximum (airborne concentration) level above which only a highly reliable breathing apparatus providing maximum worker protection is permitted [NIOSH 2004]. IDLH values are based on a 30-minute exposure duration.

Implantation: The process by which a fertilized egg implants in the uterine lining, typically several days following conception, depending on the species.

Inhalation Reference Concentration (RfC): An estimate (with uncertainty spanning perhaps an order of magnitude) of a continuous inhalation exposure for a chronic duration (up to a lifetime) of the human population (including sensitive subgroups) that is likely to be without an appreciable risk of deleterious effects during a lifetime [USEPA 2010]. It can be derived from a NOAEL, LOAEL, or benchmark concentration, with UF generally applied to reflect limitations of the data used. Generally used in USEPA non-cancer health assessments.

Internal Dose: A dose denoting the amount absorbed without respect to specific absorption barriers or exchange boundaries.

International Toxicity Estimates for Risk Database (ITER): A free Internet database of human health risk values and cancer classifications for over 600 chemicals of environmental concern, from multiple organizations worldwide (additional information available at http://www.tera.org/iter/).

Intraperitoneal: Within the peritoneal cavity (the area that contains the abdominal organs).

LC_{01}: The statistically determined concentration of a substance in the air that is estimated to cause death in 1% of the test animals.

LC_{50}: The statistically determined concentration of a substance in the air that is estimated to cause death in 50% (one half) of the test animals; median lethal concentration.

LC_{LO}: The lowest lethal concentration of a substance in the air reported to cause death, usually for a small percentage of the test animals.

LD$_{50}$: The statistically determined lethal dose of a substance that is estimated to cause death in 50% (one half) of the test animals; median lethal concentration.

LD$_{LO}$: The lowest dose of a substance that causes death, usually for a small percentage of the test animals.

LEL: The minimum concentration of a gas or vapor in air, below which propagation of a flame does not occur in the presence of an ignition source.

Lethality: Pertaining to or causing death; fatal; referring to the deaths resulting from acute toxicity studies. May also be used in lethality threshold to describe the point of sufficient substance concentration to begin to cause death.

Lowest Observed Adverse Effect Level (LOAEL): the lowest tested dose or concentration of a substance that has been reported to cause harmful (adverse) health effects in people or animals.

Malignant: A growth with a tendency to invade and destroy nearby tissue and spread to other parts of the body.

Maternal Toxicity: Adverse effects occurring in the mother during a developmental study. Maternal toxicity can result in adverse effects to the fetus.

Maximum Likelihood Concentration: A statistical estimate of the concentration that was most likely to cause the desired effect.

Mode of Action: The sequence of significant events and processes that describe how a substance causes a toxic outcome. Mode of action is distinguished from the more detailed mechanism of action, which implies a more detailed understanding on a molecular level.

Nominal Concentration: The concentration of test article introduced into a chamber. It is calculated by dividing the mass of test article generated by the volume of air passed through the chamber. The nominal concentration does not necessarily reflect the concentration to which an animal is exposed.

No Observed Adverse Effect Level (NOAEL): The highest tested dose or concentration of a substance that has been reported to cause no harmful (adverse) health effects in people or animals.

Occupational Exposure Limit (OEL): Workplace exposure recommendations developed by governmental agencies and non-govermental organizations. OELs are intended to represent the maximum airborne concentrations of a chemical substance below which workplace exposures should not cause adverse health effects. OELs may apply to ceiling, short-term (STELs), or time-weighted average (TWA) limits.

Parturition: The act or process of giving birth.

Peak Concentration: Highest concentration of a substance recorded during a certain period of observation.

Permissible Exposure Limit (PEL): Occupational exposure limits developed by OSHA (29 CFR 1910.1000) or MSHA (30 CFR 57.5001) for allowable occupational airborne exposure concentrations. PELs are legally enforceable and may be designated as ceiling, STEL, or TWA limits.

Permit-Required Confined Spaces: OSHA defines a confined space as one that has one or more of the following characteristics: (1) contains or has the potential to contain a hazardous atmosphere; (2) contains a material that has the potential to engulf an entrant; (3) has walls that converge inward or floors that slope downward and taper into a smaller area which could trap or asphyxiate an entrant; (4) or contains any other recognized safety or health hazard, such as unguarded machinery, exposed live wires, or heat stress.

Physiologically Based Pharmacokinetic (PBPK) Model: A model that estimates the dose to a target tissue or organ by taking into account the rate of absorption into the body, distribution among target organs and tissues, metabolism, and excretion.

Point of Departure (POD): The point on the dose–response curve from which dose extrapolation is initiated. This point can be the lower bound on dose for an estimated incidence or a change in response level from a concentration-response model (BMC), or it can be a NOAEL or LOAEL for an observed effect selected from a dose evaluated in a health effects or toxicology study.

Promulgation: To make known (a decree, for example) by public declaration; announce officially.

Provisional Advisory Level (PAL): A tiered set of air and drinking water threshold exposure values for high priority chemical, biological, and radiological agents intended for the general public, including susceptible and sensitive subpopulations. Developed by USEPA to inform risk-based decision-making during a response to terrorist or natural disaster incidents [US DHS 2009].

RD_{50}: The statistically determined concentration of a substance in the air that is estimated to cause a 50% (one half) decrease in the respiratory rate.

Recommended Exposure Limit (REL): Recommended maximum exposure limit to prevent adverse health effects based on human and animal studies and established for occupational (up to 10-hour shift, 40-hour week) inhalation exposure by NIOSH. RELs may be designated as ceiling, STEL, or TWA limits.

Reproductive Toxicology: The study of adverse effects on male and/or female reproductive function, capacity, or associated endocrine system components. Common adverse effects include altered sexual behavior, fertility, pregnancy outcomes, or modifications in other functions that depend on reproductive integrity of the system.

Risk Phrases: A European system of hazard codes and phrases for labeling dangerous substances and compounds, consisting of the letter R followed by a series of numbers. Each number corresponds to a specific hazard phrase. For example, R-34 means "causes burns," regardless of any language translations.

Sensory Irritation: Immediate irritation to the eyes and nose, due to an interaction between the substance and receptors in the trigeminal nerve endings. Often an endpoint for OEL derivation.

Short-Term Exposure: Repeated exposure by the oral, dermal, or inhalation route for more than 24 hours, up to 30 days.

Short-Term Exposure Limit (STEL): A worker's 15-minute time-weighted average exposure concentration that shall not be exceeded at any time during a work day.

Short-Term Public Emergency Guidance Level (SPEGL): A ceiling guidance level for unpredicted, single, short-term, emergency exposures (1 to 24 hours) for the general public. SPEGLs are developed at the request of the U.S. Department of Defense by the National Research Council's Committee on Toxicology [NAS 1986].

Spacecraft Maximum Allowable Concentration (SMAC): Guideline values set to protect astronauts from spacecraft contaminants. Short-term guidelines (1 to 24 hours) apply to accidental releases, and long-term guidelines (up to 180 days) apply to low levels of contaminants aboard a spacecraft. These guidelines are set by the NASA/JSC in cooperation with the National Research Council's Committee on Toxicology [NASA 1999].

Surrogate: Relatively well studied chemical whose properties are assumed, with appropriate adjustments for differences in potency, to apply to an entire chemically and toxicologically related class; for example, benzo(*a*)pyrene data are assumed to be toxicologically equivalent to those for all carcinogenic polynuclear aromatic hydrocarbons or are used as a basis for extrapolating to these other chemicals.

Systemic Concentration: The concentration in a blood or tissue arising from exposure to a substance that is absorbed and distributed throughout the body.

Target Organ: Organ in which the toxic injury manifests in terms of dysfunction or overt disease.

Temporary Emergency Exposure Limits (TEELs): Tiered temporary guidance values that are used by DOE until AEGL or ERPG values are available. TEELs are derived by the Subcommittee on Consequence Assessment and Protective Actions (SCAPA) to aid in emergency preparedness hazard analysis of DOE facilities, employees, and adjacent communities in the event of an accidental chemical release [US DOE 2008].

Threshold Limit Values (TLVs®): Recommended guidelines for occupational exposure to airborne contaminants, published by the American Conference of Governmental Industrial Hygienists (ACGIH). TLVs refer to airborne concentrations of chemical substances and represent conditions under which it is believed that nearly all workers may be repeatedly exposed, day after day, over a working lifetime, without adverse effects. TLVs may be designated as ceiling, short-term (STELs), or 8-hr TWA limits.

Time-Weighted Average (TWA): A worker's 8-hour (or up to 10-hour) time-weighted average exposure concentration that shall not be exceeded during an 8-hour (or up to 10-hour) work shift of a 40-hour week. The average concentration is weighted to take into account the duration of different exposure concentrations.

Toxic Inhalation Hazard (TIH): Gases or volatile liquids that are known or presumed on the basis of tests to be so toxic to humans as to pose a hazard to health in the event of a release during transportation, determined by DOT.

Toxicity: The degree to which a substance is able to cause an adverse effect on an exposed organism.

Toxicology: Scientific discipline involving the study of the actual or potential danger presented by the harmful effects of substances (poisons) on living organisms and ecosystems, of the relationship of such harmful effects to exposure, and of the mechanisms of action, diagnosis, prevention, and treatment of intoxications.

Tumor: An abnormal mass of tissue that results from excessive cell division that is uncontrolled and progressive. Tumors perform no useful body function. Tumors can be either benign (not cancerous) or malignant (cancerous).

Uncertainty Factors: Mathematical adjustments applied to the POD when developing IDLH values. The UFs for IDLH value derivation are determined by considering the study and effect used for the POD, with further modification based on the overall database.

Weight of Evidence (Toxicity): Extent to which the available biomedical data support a conclusion, such as whether a substance causes a defined toxic effect (e.g., cancer in humans), or whether an effect occurs at a specific exposure level.

Workplace Environmental Exposure Levels (WEELs): Exposure levels that provide guidance for protecting most workers from adverse health effects related to occupational chemical exposures expressed as a TWA or ceiling limit.

Acknowledgements

This document was developed by the Education and Information Division (Paul Schulte, Ph.D., Director). G. Scott Dotson, Ph.D., was the project officer and lead NIOSH author for this CIB, assisted in great part by Richard Niemeier, Ph.D. The basis for this document was a report contracted by NIOSH and prepared by Andrew Maier, Ph.D., Ann Parker, and Lynne Haber, Ph.D. (*Toxicology Excellence for Risk Assessment* [TERA]).

For their contribution to the technical content and review of this document, special acknowledgment is given to the following NIOSH personnel:

Division of Applied Research and Technology

Leonid A. Turkevich, Ph.D.

Division of Respiratory Disease Studies

Chris Coffey, Ph.D.

Division of Survelliance, Hazard Evaluations, and Field Studies

Dave Sylvain
Doug Trout, M.D.

Education and Information Division

David Dankovic, Ph.D.
Charles L. Geraci, Ph.D.
Thomas J. Lentz, Ph.D.
Chris Sofge, Ph.D.
Ralph Zumwalde, M.Sc.

National Personal Protective Technology Laboratory

Heinz Ahlers, M.Sc., J.D.
Leslie Boord, Ph.D.

NIOSH Office of the Director

K. Ann Berry, Ph.D., M.B.A.
A. Yvonne Boudreau, M.D.
Lisa Delaney, M.Sc.
Ted Katz, M.P.A.

Office of Mine Safety and Health Research

Pamela Drake, M.P.H.

Additional NIOSH contributors to the document include:

Devin S. Baker, Seleen Collins, Sarah Earl, John Lechiter, Rachel Thieman, and Vanessa B. Williams.

In addition, special appreciation is expressed to the following individuals for serving as independent, external reviewers and providing critical comments that contributed to the refinement of the document:

Rebecca Adams
Army Institute of Public Health (AIPA; formerly US Army Center
  for Health Promotion and Preventive Medicine)
US Army Public Health Command
Aberdeen Proving Grounds, Maryland

Christopher Carroll M.S.E.S., AIPA
US Army Public Health Command
Aberdeen Proving Grounds, Maryland

Ernest V. Falke Ph.D.
US Environmental Protection Agency (USEPA)
Acute Exposure Guideline Levels Program
Washington, District of Columbia

Veronique Hauschild M.P.H., AIPA
US Army Public Health Command
Aberdeen Proving Grounds, Maryland

James Holler, Ph.D.
Agency for Toxic Substances and Disease Registry (ATSDR)
Division of Toxicology and Environmental Medicine, Prevention,
  Response, and Medical Support Branch
Atlanta, Georgia

Glenn Millner, Ph.D.
Center for Toxicology and Environmental Health (CTEH), LLC
Little Rock, Arkansas

John S. Morawetz, M.Sc.
ICWUC Center for Worker Health & Safety Education
Cincinnati, Ohio

Mattias Öberg, Ph.D.
Unit for Work Environment Health
Institute of Environmental Medicine, Karolinska Institutet
Stockholm, Sweden

Irene L. Richardson, M.Sc., R.S., AIPA
US Army Public Health Command
Aberdeen Proving Grounds, Maryland

Laurie Roszell, Ph.D., AIPA
US Army Public Health Command
Aberdeen Proving Grounds, Maryland

John J. Resta, M.C.E., AIPA
US Army Public Health Command
Aberdeen Proving Grounds, Maryland

George Rusch, Ph.D., DABT, ATS
Risk Assessment and Toxicology Services
Bridgewater, New Jersey

Richard B. Schlesinger, Ph.D.
Dyson College of Arts and Sciences
Pace University
New York

Public Review

NIOSH greatly appreciates the public comments on the December 2010 draft document that were submitted to the NIOSH docket. The comments and responses to them can be seen at: http://www.cdc.gov/niosh/docket/archive/docket156.html

1 Introduction

Occupational exposures to chemicals have long been recognized as having the potential to adversely affect the lives and health of workers. Acute or short-term exposures to high concentrations of some airborne chemicals have the ability to quickly overwhelm workers, resulting in a wide spectrum of undesirable health outcomes that may include irritation of the eyes and respiratory tract, severe irreversible health effects, impairment of the ability to escape from the exposure environment, and, in extreme cases, death. Airborne concentrations of chemicals capable of causing such adverse health effects or impeding escape from "high risk" situations or conditions may arise from a variety of situations affecting workers, including special work procedures (e.g., in confined spaces), industrial accidents (e.g., chemical spills or explosions), or chemical releases into the community (e.g., during transportation incidents or other uncontrolled release scenarios). Many organizations develop acute inhalation limits or guidelines. These are typically presented as airborne concentrations. However, airborne chemicals can also contribute to toxicity through other exposure routes, such as the skin and eyes.

The "immediately dangerous to life or health air concentration values (IDLH values)" developed by the National Institute for Occupational Safety and Health (NIOSH) characterize these high-risk exposure concentrations and conditions and are used as a component of the respirator selection criteria first developed in the mid-1970s [NIOSH 1994]. Since the development of the original IDLH values in the 1970s and their subsequent revision in 1994, NIOSH has continued to review relevant scientific data and conduct research on methods for developing acute exposure guidelines. This document reflects continuing enhancements in risk assessment approaches and provides a detailed description of the methodology used to derive IDLH values. The documentation for specific IDLH values is available as separate NIOSH publications and on the NIOSH website (http://www.cdc.gov/niosh/idlh/default.html).

The primary objectives of this Current Intelligence Bulletin (CIB) are:

1. To provide a brief history of the development of IDLH values,

2. To update the scientific bases and risk assessment methodology used to derive IDLH values from quality toxicity and human health effects data,

3. To provide transparency behind the rationale and derivation process for IDLH values, and

4. To demonstrate how scientifically credible IDLH values can be derived from available data resources.

1.1 Background

The concept of using respirators to protect workers in situations that are immediately dangerous to life or health was discussed at least as early as the 1940s. The following is from a 1944 U.S. Department of Labor (DOL) bulletin:

> The situations for which respiratory protection is required may be designated as, (1) nonemergency and (2) emergency. Nonemergency situations are the more or less normal ones that involve exposure to atmospheres that are not immediately dangerous to health and life, but will produce marked discomfort, sickness, permanent harm, or death after a prolonged exposure or with repeated exposure. Emergency situations are those that involve actual or potential exposure to atmospheres that are immediately harmful and dangerous to health or life after comparatively short exposures. [Yant 1944]

The Occupational Safety and Health Administration (OSHA) defines an IDLH concentration in the hazardous waste operations and emergency response regulation as follows:

> An atmospheric concentration of any toxic, corrosive or asphyxiant substance that poses an immediate threat to life or would interfere with an individual's ability to escape from a dangerous atmosphere [29 CFR 1910.120].

In the OSHA regulation on "permit-required for confined spaces," an IDLH condition is defined as follows:

> Any condition that poses an immediate or delayed threat to life or that would cause irreversible adverse health effects or that would interfere with an individual's ability to escape unaided from a permit space [29 CFR 1910.146]. **Note:** Some materials (e.g., hydrogen fluoride gas and cadmium vapor) may produce immediate transient effects that, even if severe, may pass without medical attention, but are followed by sudden, possibly fatal collapse ~ 6 to 24 hours after exposure. The victim "feels normal" from recovery from transient effects until collapse. Such materials in hazardous quantities are considered to be "immediately dangerous to life or health." [29 CFR 1910.146]

In the current respiratory protection standard, OSHA states that an IDLH condition is as follows:

> An atmosphere that poses an immediate threat to life, would cause irreversible adverse health effects, or would impair an individual's ability to escape from a dangerous atmosphere [29 CFR 1910.134].

As part of this standard, additional guidance is provided by OSHA that dictates the type and application of respirators in IDLH conditions. Specific information that is provided in the respiratory protection standard requires:

- A trained standby person be present with suitable rescue equipment when self-contained breathing apparatus or hose masks with blowers are used in IDLH atmospheres; and

- Persons using air-line respirators in IDLH atmospheres must be equipped with safety harnesses and safety lines for lifting or removing workers from hazardous atmospheres.

The Mine Safety and Health Administration (MSHA) defines IDLH as "immediately harmful to life" [30 CFR 56/57/5005(c)]. The standard defines "immediately harmful to life" as that used by NIOSH to define "immediately dangerous to life or health," which is "acute respiratory exposure that poses an immediate threat of loss of life, immediate or delayed irreversible adverse health effects, or acute eye exposure that would prevent escape from a hazardous atmosphere." IDLH values are based on a 30-minute exposure duration.

1.2 The Standards Completion Program

In 1974, NIOSH and OSHA jointly initiated the development of occupational health standards consistent with Section 6(b) of the Occupational Safety and Health Act of 1970 for substances with then-existing OSHA permissible exposure limits (PELs). This joint effort was called the Standards Completion Program (SCP) and resulted in the development of 387 substance-specific draft standards with supporting documentation that contained technical information and recommendations needed for the promulgation of occupational health regulations. Although standards were not promulgated at that time, these data became the original basis for the *NIOSH/OSHA Occupational Health Guidelines for Chemical Hazards* [NIOSH/OSHA 1981].

As part of the respirator selection process for each draft technical standard, an IDLH value was determined for each chemical. The definition used for IDLH values that was derived during the SCP was based on the definition stipulated in 30 CFR 11.3(t). The purpose of deriving an IDLH value was to provide guidance on respirator selection and to establish a maximum exposure concentration in which workers, in the event of respiratory protection failure (e.g., contaminant breakthrough in a cartridge respirator or stoppage of air flow in a supplied-air respirator), could escape safely when the exposure was below the IDLH value. In determining IDLH values, the ability of a worker to escape without loss of life or irreversible health effects was considered, along with severe eye or respiratory tract irritation and other deleterious effects (e.g., disorientation or incoordination) that could prevent escape. Although in most cases, egress from a particular worksite could occur in much less than 30 minutes, as a safety margin, IDLH values were based on the effects that might occur as a consequence of a 30-minute exposure. However, the 30-minute period was NOT meant to imply that workers should stay in the work environment any longer than necessary following the failure of respiratory protection equipment; in fact, **EVERY EFFORT SHOULD BE MADE TO EXIT IMMEDIATELY!**

1.3 Basis of the Original IDLH Values

IDLH values were determined for each substance during the SCP on a case-by-case basis, taking into account the toxicity data available at the time. Whenever possible, IDLH values were determined with use of health effects data from studies of humans exposed for short durations. However, in most instances, a lack of human data necessitated the use of animal toxicity data. When the findings of inhalation studies of animals exposed for short durations (i.e., 30 minutes to 4 hours) were the only health effects data available, IDLH values were based on the lowest exposure causing death or irreversible health effects in any species. When lethal dose (LD) data from animals were used, IDLH values were estimated on the basis of an equivalent exposure to a 70-kilogram (kg) worker breathing 10 cubic meters (m^3) of air for an 8-hour period. Because chronic exposure data may have little relevance to acute effects, these types of data were used in determining IDLH values only when no acute toxicity data were available and only in conjunction with competent scientific judgment. In a number of instances when no relevant human or animal toxicity data were available, IDLH values were based on analogies with other substances with similar toxic effects.

The basis for each of the 387 original IDLH values determined during the SCP was reviewed and paraphrased from the individual draft technical standards for the publication of the original list of IDLH values. Also included is a complete listing of references cited in the SCP; in many cases where only secondary references were cited, the original sources have also been added. Whenever available, the references (secondary and primary) were obtained to verify the information cited in the SCP. However, a few of the original references, such as personal communications and foreign reports, could not be located.

Although 387 substances were originally included in the SCP, IDLH values were not determined for all of them. The published data at that time for 40 of these substances—for example, DDT (Chemical Abstracts Service number [CAS#] 50-29-3) and triphenyl phosphate (CAS# 115-86-6)—showed no evidence that an acute exposure to high concentrations would impede escape or cause any irreversible health effects following a 30-minute exposure, and the designation "NO EVIDENCE" was used in the listing of IDLH values. For all of these substances, respirators were selected on the basis of assigned protection factors. For some (e.g., copper fume (CAS# 7440-50-8) and tetryl (CAS #479-45-8), an assigned protection factor of 2,000 times the PEL was used to determine the concentration above which only the "most protective" respirators were permitted. However, for most particulate substances for which evidence for establishing an IDLH value did not exist (e.g., ferbam [CAS# 14484-64-1] and oil mist [CAS# 8012-95-1]), the use of an assigned protection factor of 2,000 would have resulted in the assignment of respirators at concentrations that were not likely to be encountered in the occupational environment. In addition, exposure concentrations greater than 500 times the PEL for many airborne particulates could result in exposures that would hamper vision. Therefore, it was decided as part of the SCP (and during the review and revision of the IDLH values) that for such particulate substances, only the "most protective" respirators would be permitted for use in concentrations exceeding 500 times the PEL.

IDLH values could not be determined during the SCP for 22 substances (e.g., bromoform [CAS# 75-25-2] and calcium oxide [CAS# 1305-78-8]) because of a lack of relevant toxicity data, and therefore, the designation "UNKNOWN" was used in the IDLH value listing. For most of these substances, the concentrations above which only the "most protective" respirators were allowed were based on assigned protection factors that ranged from 10 to 2,000 times the PEL, depending on the substance. There were also 10 substances (e.g., n-pentane [CAS# 109-66-0] and ethyl ether [CAS# 60-29-7]) for which it was determined only that the IDLH values were in excess of the lower explosive limits (LELs). Therefore, the LEL was selected as the IDLH value, with the designation "LEL" added in the IDLH value listing. For these substances, only the "most protective" respirators were permitted above the LEL in the SCP draft technical standards.

For 14 substances (e.g., beryllium [CAS# 7440-41-7] and endrin [CAS# 72-20-8]), the IDLH values determined during the SCP were greater than the concentrations permitted on the basis of assigned respiratory protection factors. In most instances the IDLH values for these substances were set at concentrations 2,000 times the PEL.

1.4 Update of the IDLH Values in 1994

The NIOSH definition for an IDLH condition, as given in the *NIOSH Respirator Decision Logic* [NIOSH 2004], is a situation "that poses a threat of exposure to airborne contaminants when that exposure is likely to cause death or immediate or delayed permanent adverse health effects or prevent escape from such an environment." It is also stated that the purpose of establishing an IDLH value is to "ensure that the worker can escape from a given contaminated environment in the event of failure of the respiratory protection equipment."

The respirator decision logic uses an IDLH value as one of several respirator selection criteria. "Highly reliable" respirators (i.e., the most protective respirators) would be selected for emergency situations, firefighting, exposure to carcinogens, entry into oxygen-deficient atmospheres, entry into atmospheres that contain a substance at a concentration greater than 2,000 times the NIOSH recommended exposure limit (REL) or OSHA PEL, and entry into IDLH conditions. These "highly reliable" respirators include either a self-contained breathing apparatus (SCBA) that has a full face piece and is operated in a pressure-demand or other positive-pressure mode or else a supplied-air respirator that has a full face piece and is operated in a pressure-demand or other positive-pressure mode in combination with an auxiliary SCBA operated in a pressure-demand or other positive-pressure mode.

When the IDLH values were developed in the mid-1970s, only limited toxicological data were available for many of the substances. In 1993, NIOSH requested information on the uses of IDLH values in the workplace and on the scientific adequacy of the criteria and procedures originally used for establishing them [Federal Register, Volume 58, Number 229, p. 63379, Wednesday, December 1, 1993]. The information received in response to the Federal Register announcement was evaluated and used to establish future actions concerning IDLH values.

While new methodology research efforts were planned and initiated, NIOSH also decided to review the original IDLH values and revise them as appropriate [NIOSH 1994]. The update was completed in 1994. The 1994 update also included revisions or derivation of IDLH values for 85 substances (e.g., benzene [CAS# 71-43-2] and methylene chloride [CAS# 75-09-2]) determined by NIOSH to meet the OSHA definition of "potential occupational carcinogen," as given in 29 CFR 1990.103. For all of these substances, except ethylene oxide (CAS#75-21-8) and crystalline silica (CAS# 14808-60-7), NIOSH recommends that the "most protective" respirators be worn by workers exposed at concentrations above the NIOSH REL, or at any detectable concentration when there is no REL. For ethylene oxide and crystalline silica, NIOSH recommends that the "most protective" respirators be worn in concentrations exceeding 5 parts per million (ppm) and milligrams per cubic meter of air (mg/m^3), respectively [NIOSH 1987, 2004].

1.5 Purpose and Objectives of the IDLH Values

IDLH values have traditionally been identified as a key component of the decision logic for the selection of respiratory protection devices. For example, the *NIOSH Respirator Selection Logic* [NIOSH 2004] states that the purpose of establishing an IDLH value is (1) to ensure that the worker can escape from a given contaminated environment in the event of failure of the respiratory protection equipment and (2) to determine a maximum level above which only a highly reliable breathing apparatus providing maximum worker protection is permitted. Additionally, IDLH values are based on a 30-minute exposure duration. Since the inception of IDLH values as part of the SCP, occupational health professionals have employed these values beyond their initial purpose as a component of the *NIOSH Respirator Selection Logic*. Examples of such applications of the IDLH values include the development of Risk Management Plans (RMPs) for non-routine work practices governing operations in high-risk environments (e.g., confined spaces) and the development of Emergency Preparedness Plans (EPPs), which provide guidance for emergency response personnel and workers during unplanned exposure events. This CIB presents the methodology used to derive IDLH values capable of being used in both the traditional role of respirator selection and in non-traditional applications, including the development of RMPs and EPPs.

The scientific rationale and derivation process outlined in this CIB have been established to ensure that a consistent approach is used for development of IDLH values. According to this protocol, IDLH values are based on health effects considerations determined through a critical assessment of the toxicology and human health effects data. This approach ensures that the IDLH values reflect an airborne concentration of a substance that represents a high-risk situation that may endanger workers' life or health. The emphasis on health effects is consistent with both the traditional use of IDLH values as a component of the respirator selection logic and the growing applications of IDLH values in guiding accident prevention and emergency response planning. It is important to note that IDLH values are concentrations that may cause adverse effects, and thus, they are not intended to be used as surrogates for occupational exposure limits (OELs). OELs, such as NIOSH RELs, are intended to protect workers from adverse health effects associated with repeated chemical exposure for up to 10-hour shifts during a 40-hour work week for a working lifetime. The IDLH values should not be used as comparative indices of toxicity or to infer a "safe" level for exposures to chemicals under routine occupational exposure conditions (see *Section 2.3*). A situation resulting in airborne concentrations at or near the IDLH value should be considered a non-routine event, and exposure duration should not exceed 30 minutes. All available precautions should be taken to ensure that workers exit the environment immediately if exposures are at or near concentrations equivalent to IDLH values.

NIOSH recognizes that in some cases a health-based IDLH value might not account for all workplace hazards, such as safety concerns and considerations. Situations and conditions that might preclude the use of a health-based IDLH value include but are not limited to these:

- Where the IDLH value based on health effects considerations is above the concentration that would result in oxygen deprivation (oxygen concentration of less than 19.5%). Chemicals capable of causing such conditions include inert gases such as argon (CAS# 7440–37–1), carbon dioxide (CAS# 124-38-9), and nitrogen (CAS# 7727-37-9).

- Where the IDLH value based on health effects considerations is higher than a particulate concentration that generates significant hazards from reduced visibility. Such conditions may occur in processes that generate dust plumes in enclosed areas or confined spaces (e.g., grinding, milling, or mining operations) and structural fires.

- Where the IDLH value based on health-effects considerations is greater than 10% of the LEL concentration or the minimum concentration of gas or vapor in air below which propagation of a flame does not occur in the presence of an ignition source. Chemicals capable of causing such conditions include flammable gases or vapors such as acetone (CAS# 67-64-1), ethyl acetate (CAS# 64-17-5), and n-pentane (CAS#109-66-0).

- Where the IDLH value based on health effects considerations is greater than the time-weighted average (TWA) OEL multiplied by the assigned protection factor for the most protective respirator. Because IDLH values are based on acute exposure and health effects data, the most protective respirator may not be adequately protective for full-shift exposures at this concentration. Examples of substances where this situation may occur include chromic acid and chromates (CAS# 1333-82-0) and lead compounds (CAS# 7439-92-1, metal).

In such cases, it is important that safety hazards or other considerations be taken into account. Information on safety hazards will be incorporated into the derivation of an IDLH value when appropriate. For example, in the event that the derived health-based IDLH value exceeds 10% of the LEL concentration for a flammable gas or vapor, the airborne concentration that is equal to 10% of the LEL will become the IDLH value for the chemical. The following hazard statement will be included in the support documentation: "The health-based IDLH value is greater than 10% of the LEL (>10% LEL) of the chemical of interest in the air. Safety considerations related to the potential hazard of explosion must be taken into account." In addition, the notation (>10% LEL) will appear beside the IDLH value in the *NIOSH Pocket Guide to Chemical Hazards* [NIOSH 2005] and other NIOSH publications. Similar statements will be developed as needed for other safety hazards and considerations. The use of hazard statements and notations to provide supplemental information on safety hazards and considerations aligns with the protocols used to derive the AEGLs by the National Advisory Committee for Acute Exposure Guideline Levels for Hazardous Substances (NAC/AEGL Committee) [NAS 2001]. Additional information on the establishment of IDLH values based on safety hazards can be located in Section 3.6.

2 Comparison of IDLH Values to Alternative Short-term Exposure Limits/Values

An important step in the development of IDLH values is the review of alternative short-term exposure limits/values developed by other agencies and organizations. The review of such information serves several purposes:

- It is useful for verifying that all key data and scientific issues are considered and thus serves as one step in verifying that a robust literature search has been completed.

- It assists in identifying critical issues with study design, methodology, or results for critical studies that must be considered in developing an IDLH value.

- In some cases, alternative exposure limits/values may aid in determining a potential range for the IDLH value (after taking into account the methodology differences used to develop various short-term limits/values), as described later in this section.

Because the documentation for the IDLH values is intended to be a concise summary document, NIOSH incorporates in the IDLH documentation information on the acute effects of chemicals and selected short-term limits/values from other in-depth peer-reviewed assessments, for comparison purposes. Table 2–1 summarizes several of the short-term exposure limits/values most commonly evaluated during the derivation of IDLH values. There are numerous other sources of short-term exposure limits/values, which may be reviewed on a case-by-case basis for a particular chemical, depending on availability.

Although IDLH values may rely on much of the same acute health effects information used to derive alternative short-term exposure limits/values, there are underlying differences in the intended use of the various acute exposure values. Therefore, review of documentation for these alternative short-term limits/values provides information to guide IDLH value development, but the actual proposed values are not directly comparable. The remaining sections of Chapter 2 discuss the different purposes and populations protected by commonly reviewed alternative short-term exposure limits/values.

2.1 Acute Exposure Guideline Levels

AEGLs are threshold exposure limits for the general public intended to be guideline levels used during rare events or single once-in-a-lifetime exposures to airborne concentrations of acutely toxic, high-priority chemicals [NAS 2001]. The threshold exposure limits are designed to protect the general population, including the elderly, children or other potentially sensitive groups that are generally not considered in the development of workplace exposure recommendations [NAS 2001]. AEGLs are based primarily on acute toxicology data and not subchronic or chronic data and therefore do not reflect the health effects that could result from frequent exposures.

Three levels, referred to as AEGL-1, AEGL-2, and AEGL-3, are developed for each of five exposure periods (10 minutes, 30 minutes, 1 hour, 4 hours, and 8 hours) and are distinguished by varying degrees of severity of toxic effects. The three AEGLs are defined as follows [NAS 2001]:

- **AEGL-1** is the airborne concentration (expressed as ppm or mg/m^3) of a substance above which it is predicted that the general

Table 2–1. Short-term exposure limits/values by other agencies and organizations

Purpose of short-term exposure limit	Agency or organization designation
Acute exposure guidelines for protection of the general public during emergency or rare releases	Acute Exposure Guidelines Levels (AEGLs)
	Emergency Response Planning Guidelines (ERPGs)
	Other values as appropriate
Acute exposure guidelines for potential routine acute exposures in the workplace such as short-term exposure limits (STELs) or Ceiling Limits ("C").	National Institute for Occupational Safety and Health (NIOSH) Recommended Exposure Limits (RELs)
	Occupational Safety and Health Administration (OSHA) Permissible Exposure Limits (PELs)
	American Conference of Governmental Industrial Hygienists (ACGIH) Threshold Limit Values (TLVs)®
	American Industrial Hygiene Association (AIHA) Workplace Environmental Exposure Levels (WEELs)
	Other values as appropriate

population, including susceptible individuals, could experience notable discomfort, irritation, or certain asymptomatic, non-sensory effects. However, the effects are not disabling and are transient and reversible upon cessation of exposure.

- **AEGL-2** is the airborne concentration (expressed as ppm or mg/m^3) of a substance above which it is predicted that the general population, including susceptible individuals, could experience irreversible or other serious, long-lasting adverse health effects or an impaired ability to escape.

- **AEGL-3** is the airborne concentration (expressed as ppm or mg/m^3) of a substance above which it is predicted that the general population, including susceptible individuals, could experience life-threatening health effects or death.

Airborne concentrations below the AEGL-1 represent exposure levels that could produce mild and progressively increasing irritation or asymptomatic, non-sensory effects, such as non-disabling odor and taste. With increasing airborne concentrations above each AEGL, there is a progressive increase in the likelihood of occurrence and the severity of effects described for each corresponding AEGL. Although the AEGL values represent threshold levels for the general public, including susceptible subpopulations, such as infants, children, the elderly, persons with asthma, and those with other illnesses, it is recognized that individuals, subject to unique or idiosyncratic responses, could experience the effects described at concentrations below the corresponding AEGL.

Like the IDLH value, the AEGL-2 is designed to protect from irreversible or other serious effects and escape-impairing effects. Thus, the effects that are the basis for the AEGL-2 closely match those of interest for the IDLH value. In addition, the AEGLs include a 30-minute value, which is the same duration of interest for the IDLH values. One significant difference between the IDLH value and the AEGL-2 is that the AEGL-2 is designed to protect the general population, including potentially sensitive subpopulations (i.e., children, elderly, and individuals with pre-existing health impairments). IDLH values are designed for worker populations, which traditionally exclude the most sensitive subpopulations. This assumption is based on the consideration that there would be a smaller likelihood for significant inclusion of specific sensitive subpopulations in the population of working adults. In addition, the selection of the critical effect (health endpoint) and interpretation of the severity of the health impact to the population of interest (in this case a worker population in a high-risk environment) may be different than that used for the AEGL-2. This means that given the same set of data, the IDLH value will often be in the range of the 30-minute AEGL-2 but will vary somewhat because of the fundamental differences between the approaches applied to establish AEGL values and IDLH values for a chemical. The IDLH value is usually below the 30-minute AEGL-3, since, for most chemicals, serious or escape-impairing effects relevant for IDLH values occur at concentrations below the lethality threshold. In light of these considerations, recent AEGL-2 and AEGL-3 values can provide a rough gauge for identifying a potential range for the IDLH value. Exceptions may occur, partially because the AEGL process follows fairly strict methodology guidelines [NAS 2001], including the use of default approaches in the absence of chemical-specific data, whereas the process for developing IDLH values relies heavily on the overall weight of evidence, with limited use of default procedures. The extensive AEGL documentation for each chemical has been thoroughly reviewed by expert committees and is often a useful resource for de novo analyses. In addition, the AEGL documentation includes detailed analysis of all key studies, often including calculation of the value of the ten Berge exponent n [ten Berge et al. 1986]; for a detailed description of the ten Berge exponent, see *Section 3.5—Time Scaling*.

The AEGL values are derived by the NAC/AEGL Committee, which is a Federal Advisory Committee Act (FACA) committee established to identify, review, and interpret relevant toxicologic and other scientific data and to develop AEGLs for high priority, acutely toxic chemicals (available online at: http://www.epa.gov/oppt/aegl/). The NAC/AEGL includes members from federal and international agencies (e.g., NIOSH, U.S. Environmental Protection Agency [USEPA], U.S. Department of Transportation [DOT], U.S. Department of Defense [DOD], U.S. Department of Energy [DOE], Agency for Toxic Substances and Disease Registry [ATSDR], Canadian Government, Netherlands National Institute for Public Health and the Environment [RIVM], state agencies and environmental organizations, academia, private industry, and international and nonprofit organizations). Interim AEGLs prepared by the AEGL Committee, after stakeholder comment, are reviewed by the National Academy of Sciences (NAS)/National Research Council (NRC) AEGL Committee before finalization.

2.2 Emergency Response Planning Guidelines

Emergency Response Planning Guidelines (ERPGs) are developed by the American Industrial Hygiene Association (AIHA) for emergency planning and are intended as health-based guideline concentrations for single exposures to chemicals [AIHA 2006, 2008]. These guidelines (i.e., the ERPG documents and ERPG values) are intended for use as planning tools for assessing the adequacy of accident prevention and emergency response plans, including transportation emergency planning, and for developing community emergency response plans.

As with AEGLs, there are three ERPG guidance concentration levels designed for community protection [AIHA 2006]. However, ERPGs are derived for only single-exposure durations of 1 hour. Each of the three levels is defined and briefly discussed below:

- ERPG-1: The maximum airborne concentration below which it is believed that nearly all individuals could be exposed for up to one hour without experiencing other than mild, transient adverse health effects or without perceiving a clearly defined objectionable odor.

The ERPG-1 identifies a level that does not pose a health risk to the community but that may be noticeable because of slight odor or mild irritation. In the event that a small, non-threatening release has occurred, the community could be notified that they may notice an odor or slight irritation but that concentrations are below those which could cause unacceptable health effects. For some materials, because of their properties, there may not be an ERPG-1. Such cases would include substances for which sensory perception levels are higher than the ERPG-2 level. In those cases, the ERPG-1 level would be given as "Not Appropriate." It is also possible that no valid sensory perception data are available for the chemical. In these cases, the ERPG-1 level would be given as "Insufficient Data."

- **ERPG-2**: The maximum airborne concentration below which it is believed that nearly all individuals could be exposed for up to one hour without experiencing or developing irreversible or other serious health effects or symptoms that could impair an individual's ability to take protective action.

Above ERPG-2, there may be significant adverse health effects, signs, or symptoms for some members of the community that could impair their ability to take protective action. These effects might include severe eye or respiratory irritation, muscular weakness, central nervous system (CNS) impairments, or serious adverse health effects.

- **ERPG-3**: The maximum airborne concentration below which it is believed that nearly all individuals could be exposed for up to one hour without experiencing or developing life-threatening health effects.

The ERPG-3 level is a worst-case planning level, above which there is the possibility that some members of the community may develop life-threatening health effects. This guidance level could be used to determine the airborne concentration of a chemical that could pose life-threatening consequences should an accident occur. This concentration could be used in planning stages to project possible levels in the community. Once the distance from the release to the ERPG-3 level is known, the steps to mitigate the potential for such a release can be established.

Like the IDLH value, the ERPG-2 is designed to protect from irreversible or other serious and escape-impairing effects and therefore is based on effects similar to those considered as the basis for IDLH values. Like the IDLH values, ERPGs are for acute exposure, but they are based on a 1-hour rather than 30-minute exposure. All other things being equal, this would mean that ERPG-2 values will generally be lower than the corresponding IDLH values, since the potential exposure time for the ERPG is higher. Moreover, even though ERPGs are developed by an occupational health organization, ERPGs are more like the AEGLs in that they are designed to protect the general population, and thus susceptible populations are more of a consideration for ERPGs than for IDLH values.

2.3 Occupational Exposure Limits

OELs are derived by various governmental, nongovernmental, and private organizations for application to repeated or daily worker exposure situations. For example, in the United States, OELs are developed by several organizations. Examples of such organizations and their respective OEL values include; NIOSH RELs, OSHA PELs, MSHA PELs, ACGIH TLVs®, and AIHA WEELs®. Although the exact definition varies among organizations (see *Glossary*), the general intent of OELs is to identify airborne concentrations of substances in the air to which all or nearly all workers can be exposed on a repeated basis for a working lifetime without adverse health effects. OELs are developed on the basis of available human data (such as results from epidemiologic studies or controlled human exposure studies), animal toxicologic data, or a combination of human and animal data. The health basis on which exposure limits are established may differ from substance to substance; protection against impairment of health may be a guiding factor for some, whereas reasonable freedom from irritation, narcosis, nuisance, or other forms of stress may form the basis for others. For most OELs, health impairment refers to effects that shorten life expectancy, compromise physiological function, impair the capability of resisting other toxic substances or disease processes, or impair reproductive function or developmental processes. Alternative considerations, such as technological feasibility, analytical achievability and economic impact, are often included during the establishment of an OEL based on the mandate of the organization deriving the exposure limit. For this reason, it is important to review the support documentation for any OEL to determine its basis (i.e., health endpoint versus alternative endpoint) and intended purpose.

OELs are guidelines (or regulatory standards, if mandated by OSHA and MSHA) intended for use in the practice of industrial hygiene, for the control of potential workplace hazards. OELs are not intended for use in other situations, such as the evaluation or control of ambient air pollution, or for estimating the toxic potential of continuous uninterrupted exposures or other exposure scenarios involving extended work periods, or as proof of existing disease or physical conditions. OELs neither clearly delineate between safe and dangerous concentrations nor serve as a relative index of toxicity.

There are three primary categories of OELs, each with a different exposure duration comparison. The first category defines the TWA exposure concentration for up to a 10-hour workday (NIOSH REL) or a conventional 8 hour workday (OSHA PEL, MSHA PEL, ACGIH TLV®, or AIHA WEEL) during a 40-hour work week, to which it is believed that all workers (for the REL and PEL), nearly all workers (for the TLV®), or most workers (WEEL) may be repeatedly exposed daily without adverse effects. It should be noted that because alternative considerations (i.e., technical achievability, economic impact and analytical feasibility) are often included during the derivation of an OEL they may reflect an airborne concentration of a chemical for which there is residual risk of experiencing adverse health effects for some workers. The second category of OEL, called short-term exposure limit (STEL) and designated by ST preceding the value for NIOSH RELs, is a TWA concentration that should not be exceeded during any 15-minute period of a workday. ACGIH describes the TLV-STEL as the concentration to which it is believed that workers can be exposed continuously for a short period of time without suffering from irritation, chronic or irreversible tissue damage, or narcosis of sufficient degree to increase the likelihood of accidental injury, to impair self-rescue, or to materially reduce work efficiency [ACGIH 2009]. Exposures above the TLV-TWA and up to the TLV-STEL should not be longer than 15 minutes and should not occur more than four times per day, with a minimum of 60 minutes between exposures in this range [ACGIH, 2009]. The last category of OELs, referred to as ceiling OELs and designated by ACGIH with a "C" preceding the value, are the concentrations that should not be exceeded during any part of the working exposure, unless otherwise noted [ACGIH 2009].

Like the IDLH values, OELs are aimed at worker populations, and therefore consideration of susceptible populations is of less significance than for general population values. STELs and ceiling OELs are acute exposure values, whereas the TWA OELs are for repeated, chronic exposure. STELs are for a shorter duration (15 minutes), compared with 30-minute IDLH values, and repeated exposures are permitted during the work shift at these airborne concentrations. STELs can be based on some endpoints similar to those that are of concern for IDLH values (e.g., chronic or irreversible tissue damage, narcosis that would impair self-rescue). For other endpoints, the severity for the basis of STELs may be less than that for the IDLH value. For example, mild irritation that would not be escape-impairing and mild narcosis that affects work efficiency but is not escape-impairing could

Table 2–2. Other sources of acute inhalation exposure limits/values

Governmental agencies and organizations	Acute inhalation exposure limits/values	Sources
Department of Energy (DOE)	Temporary Emergency Exposure Limits (TEELs)	Craig et al. [2000]; US DOE [2008]
State agencies (California, Texas, Minnesota, New York, New Jersey, etc.)	State Exposure Limits	MDH [2010]; TCEQ [2010]; Cal/EPA [2010]; NJ RTK [2010]
National Academy of Sciences/National Research Council (NAS/NRC)	Emergency and Continuous Exposure Guidance Levels (EEGLs)	NAS [1986, 2008]
National Academy of Sciences/National Research Council (NAS/NRC)	Short-term Public Emergency Guidance Levels (SPEGLs)	NAS [1986]
NAS/NRC	Spacecraft Maximum Allowable Concentration (SMAC)	NASA [1999]
U.S. Environmental Protection Agency (USEPA)	Acute reference concentrations (RfCs)	USEPA [2010]
USEPA's homeland security program (DHS)	Provisional Advisory Levels (PALs) for Hazardous Agents	US DHS [2009]; Young et al. [2009]

be the bases for a STEL but would be considered below the threshold of interest for an IDLH value. Thus, depending on the nature of the effect caused by the chemical, the IDLH value may or may not be comparable to a STEL value for the same substance.

2.4 Other Acute Exposure Limits/Values

A number of other governmental agencies and organizations also develop, or have developed, acute inhalation exposure limits/values intended to address various applications, exposed populations, and durations. These include acute exposure limits/values listed in Table 2–2.

Documentation for acute exposure limits/values from these selected organizations is reviewed and considered if it is deemed to provide specific insights that impact the development or interpretation of the IDLH value. For example, acute exposure limits/values from other government agencies and organizations might be included in the documentation for IDLH values if they are more recent or have unique data not available in other sources.

3 Criteria for Determining IDLH Values

A weight-of-evidence approach based on scientific judgment is used in the IDLH methodology, both for evaluating the quality and consistency of the scientific data and in extrapolating from the available data to the IDLH value. The weight-of-evidence approach refers to the critical examination of all the available data from diverse lines of evidence and deriving a scientific interpretation based on the collective body of data, including its relevance, quality, and reported results. This is in contrast to a purely hierarchical (or strength-of-evidence) approach, which would use rigid decision criteria for selecting a critical adverse effect concentration and applying default uncertainty factors (UFs) to derive the IDLH value. The documentation of the IDLH value for each chemical is not intended to be a comprehensive review of all the available studies; instead, it focuses on the key data, decisions points, and scientific rationale integrated into the overall weight of evidence applied to derive the IDLH value for a chemical of interest. An example of the documentation for development of an IDLH value is provided in Appendix A, which explains the logic and rationale behind the derivation of the IDLH values for chlorine (CAS# 7782-50-5).

Because IDLH values are often developed from limited data, the process for developing a value often applies data from multiple lines of evidence rather than a single key high-quality study. Overall, the following approach is used for deriving IDLH values:

- Critical review of human and animal toxicity data to identify potential relevant studies and characterize the various lines of evidence that can support the derivation of the IDLH value

- Application of duration adjustments to determine 30-minute-equivalent exposure concentrations, as well as other dosimetry adjustments as needed

- Application of a UF for each potential POD or critical adverse-effect concentration identified from the available studies to account for issues associated with interspecies and intraspecies differences, the severity of the observed effects (including concern about cancer or reproductive or developmental toxicity), and data quality or data insufficiencies

- Developing the final recommendation for the IDLH value from the various alternative lines of evidence, using a weight-of-evidence approach, from all of the data.

Figure 3–1 provides a detailed summary of the key steps in derivation of IDLH values.

This process (see Figure 3–1) is conceptually similar to that used in other risk assessment applications, including these steps:

- Hazard characterization
- Identification of critical effects
- Identification of a POD
- Application of an appropriate UF based on the study and POD
- Determination of the final risk value.

The use of a weight-of-evidence approach allows for the integration of all available data that may originate from different lines of evidence into the analysis and the subsequent derivation of an IDLH value. Ideally, this ensures that the analysis is not restricted to a limited dataset or a single study for a specific chemical. In particular, application of the appropriate UF to each potential POD allows for consideration of the impact of the overall dataset as well as the uncertainties associated with each potential key study in determining the final IDLH value. See Appendix A for an example of how a typical dataset is evaluated to derive an IDLH value.

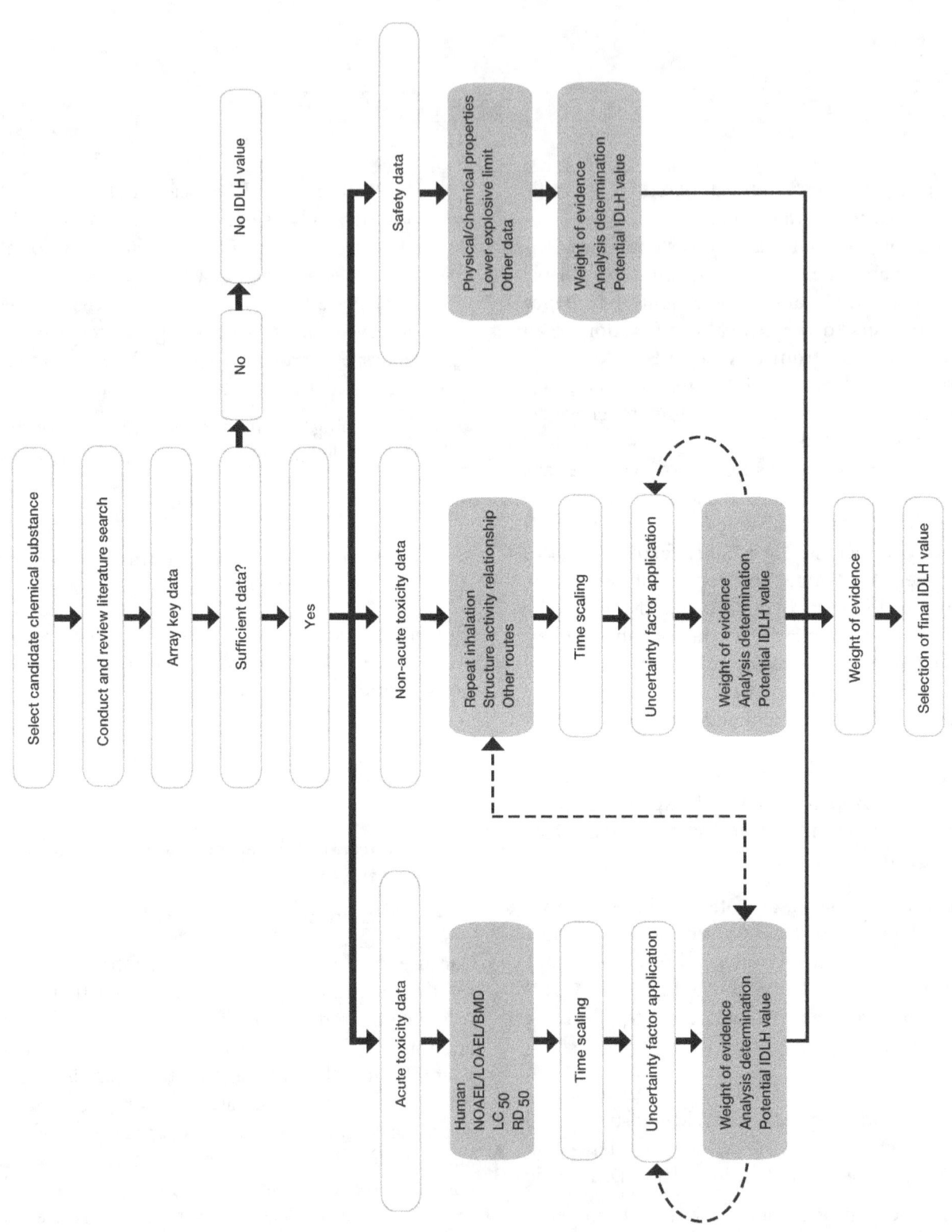

Figure 3–1. Key steps in the derivation of IDLH values.

As illustrated in the remainder of this CIB, derivation of IDLH values uses a systematic data evaluation process that gives preference for data that provide the greatest degree of confidence in the assessment. The approach describes some overall preferences that define a general data hierarchy, but the methodology allows for all of the data to be evaluated by means of a weight-of-evidence approach to develop a toxicologically meaningful IDLH value that is consistent with the dataset as a whole. Implementing such a procedure requires considerable expertise and relies heavily on weighing various lines of evidence, with vetting by multiple scientists through a rigorous peer review process. Thus, although the following sections describe general processes and priorities for use of the data, these approaches are provided as general guidance, and the focus is on interpretation of the overall database.

3.1 Importance of Mode of Action and Weight-of-Evidence Approach

The mode of action (MOA), meaning a general description of how a chemical exerts its toxic effects, is an important part of the evaluation of chemical data and development of IDLH values. MOA can be thought of as a general category of how a chemical acts to cause adverse effects. Note that the MOA is a general description of the biological basis for toxicity and does not require the detailed level of understanding implied by mechanism of action. The MOA for a chemical is identified on the basis of the observed toxic effects, any mechanistic data, structure-activity data, and information on related chemicals; many chemicals act by more than one MOA. For example, many solvents cause both respiratory irritation and CNS effects. Some of the more common classes of MOA that are encountered in developing IDLH values, and examples of chemicals that fall into these classes, include:

- **Irritants:** All chemicals in this group induce sensory irritation that is caused by trigeminal nerve stimulation and manifests as pain within the mucous membranes at the sites of contact. Chemicals such as capsaicin (CAS# 618-92-8) act in this manner without causing tissue damage. However, most sensory irritants can cause cytotoxicity (i.e., inflammation and tissue damage), with severity increasing in proportion to airborne concentrations. These types of irritants include highly reactive and/or corrosive chemicals, including acids, bases, and halogen gases. Endpoints commonly reported include eye, nose, and throat irritation, with higher concentrations typically leading to irritation and tissue damage lower in the respiratory tract. Chemicals in this class include organic solvents (e.g., vinyl acetate [CAS# 108-05-4]), organic acids (e.g., acrylic acid [CAS# 79-10-7]), halogens and other reactive gases (e.g., bromine [CAS# 7726-95-6]), and some metal compounds (e.g., titanium tetrachloride [CAS# 7550-45-0]).

- **Nervous System Effects:** Chemicals can cause nervous system effects by different MOAs. Many solvents (e.g., chloroform [CAS# 67-66-3] and 1,1,1-trichloroethane [CAS# 71-55-6]), as well as other chemicals, cause CNS depression. Clinical signs reported in humans may include fatigue, weakness, and headaches. Endpoints commonly reported in animals or humans include sedation and reduced performance in specialized neurological testing. Certain classes of pesticides (e.g., organophosphates and carbamates) and nerve agents (e.g., sarin [CAS# 107-44-8]) inhibit the action of the enzyme acetylcholinesterase. Early signs of exposures to such agents include miosis (constriction of the eye pupil), excessive salivation, and muscle twitching.

- **Metabolic Toxicants:** This class of chemicals acts by interfering with the cell's ability to generate and store energy and includes, for example, cyanides and azides. Initial effects of these chemicals are CNS symptoms (some similar to those noted previously for CNS depressants) and toxicity, ultimately leading to respiratory failure.

- **Target Organ Toxicants:** Certain organs or organ systems, such as the liver or kidney, are the site of toxicity for many chemicals. Organ-specific effects are typically not evaluated in acute lethality studies. In-depth study of a single inhalation exposure may include evaluation of histopathology or clinical chemistry for certain organ systems. Also, acute poisoning incidents in humans may indicate that the liver or kidney is a target. These organ systems are frequently the most sensitive systemic targets because of the high blood flow to these organs and their capacity for metabolizing chemicals to more reactive forms.

  In addition, some chemicals target specific organs or have unique systematic effects. For example, arsine (CAS# 7784-42-1) causes hemolysis (breakage of red blood cells), with accompanying symptoms of headache, nausea, and shortness of breath. A number of halogenated hydrocarbons (e.g., vinyl chloride [CAS# 75-01-4], HFC-134a [CAS# 811-87-2], and HCFC-141b [CAS# 1717-00-6]) cause cardiac sensitization. Chemicals, such as hexafluoroacetone (CAS# 684-16-2) and 1-bromopropane (CAS# 106-94-5), cause reproductive toxicity and development effects. Also, hormonally-mediated effects can be suggested by direct observations of effects on reproductive function or toxicity studies evaluating fetal development.

- **Asphyxiants:** Inert gases (e.g., nitrogen [CAS# 7727-37-9] and argon [CAS# 7440-37-1]) cause health effects by displacing oxygen. Chemical asphyxiants (e.g., carbon monoxide [CAS# 630-08-0], hydrogen cyanide [HCN; CAS# 74-90-8], and hydrogen sulfide [CAS# 7783-06-4]) can interfere with the body's ability to use oxygen. Some early symptoms of asphyxiation include headache, rapid breathing, heart palpitations, and lethargy.

MOA is considered as part of the evaluation of need for and adequacy of UF in extrapolation from various points of departure. The MOA of a substance is used during the derivation of IDLH values to determine UF, time extrapolation, choice of POD, and consideration of interspecies differences. Below are some examples of how MOA affects these considerations.

- A smaller UF is used when the endpoint is known to be very sensitive (e.g., cardiac sensitization in response to an epinephrine challenge, which is considered a sensitive marker of a severe effect).

- MOA information may also be used to support a flatter time extrapolation curve for sensory irritants, based on the observation that effects from such chemicals (after the first few minutes of exposure) are driven primarily by concentration and less by duration of exposure.

- MOA information indicating that the chemical targets the route of entry, with resulting effects such as eye, nose, and throat irritation, would indicate that the route-to-route extrapolation is not appropriate.

- MOA information may suggest the use of surrogates when information on the chemical of interest is limited or when a breakdown product is identified as being the primary cause of toxicity. For example, HCN is commonly used as a surrogate for acetocyanohydrin (CAS# 78-97-7), which spontaneously forms acetone and HCN. Another example is the use of hydrogen chloride (CAS# 7647-01-0) as a surrogate for chlorosilanes, which decomposes when exposed to water (i.e., humidity) to form hydrogen chloride and silanols. In both cases, the surrogates (i.e., HCN and hydrogen chloride) are directly linked to the severity of the toxic effect.

- Finally, MOA information may suggest potential refinements to the dose–response analysis. For example, carbon monoxide toxicity is due to the formation of carboxyhemoglobin (COHb), and the IDLH value for carbon monoxide is based on calculated COHb levels.

3.2 Process for Prioritization of Chemicals

In addition to serving as a crucial factor in the selection of respiratory protection equipment, IDLH values play an important role in planning work practices surrounding potential emergency high-exposure environments in the workplace and in guiding actions by emergency response personnel during unplanned exposure events. Ideally, such guidance values would be available for all chemicals that might be present under high-exposure situations. However, the development of IDLH values is not necessary for many chemicals, such as those with very low exposure potential or those that do not exhibit significant acute toxicity via the inhalation route. A prioritization process is used by NIOSH to ensure that resources allocated to IDLH value development yield the greatest impact on risk reduction. This process takes into account both toxicity and exposure potential and is applied to a broad range of potentially hazardous chemicals (e.g., chemical warfare agents, industrial chemicals, or agrochemicals) subject to emergency or uncontrolled releases. A qualitative algorithm is used to generate a priority ranking. This process provides *initial* priority rankings based on a simple approach that uses readily available sources of information. More sophisticated hazard- or risk-based ranking schemes could be used, but gathering and analyzing the data would require the same approximate effort required to actually derive an IDLH value. A complex ranking approach would not meet the primary objective to quickly and efficiently identify chemicals of greatest concern. The resulting priorities are further modified according to NIOSH emphasis areas. For example, chemicals can be added or removed from the priority list on the basis of new information related to toxicity or exposure potential. The development and use of a documented prioritization process allows for more frequent updating by NIOSH of both input data and prioritization criteria to meet changing needs. The prioritization approach is described more fully in Appendix B.

3.3 Literature Search Strategy

NIOSH performs in-depth literature searches to ensure that all relevant data from human and animal studies with acute exposures to the substance are identified. An initial literature search is done, including searches for information from the sources listed in Table 3–1.

Electronic searches of these databases are conducted with limitations on search dates. The databases are searched for studies pertinent to acute inhalation toxicity, with use of the search terms summarized in Table 3–2.

The electronic literature searches are screened for relevant articles, and a bibliography of relevant literature is compiled that identifies studies for retrieval and review. Peer-reviewed toxicology reviews are also examined, including those identified by searching the databases and organization websites, as noted in Table 3–1. Toxicology reviews that are routinely used to identify pertinent literature for developing the IDLH value include those published by ACGIH (i.e., TLV® and Biological Exposure Indices), AIHA (i.e., ERPG and WEEL documentation), ATSDR (i.e., Toxicology Profiles), National Toxicology Program (NTP), NIOSH (i.e., REL documentation), NRC (i.e., AEGL documentation), OSHA (i.e., PEL documentation), WHO (i.e., Environmental Health Criteria) and USEPA (i.e., IRIS Toxicological Reviews). Other key unpublished literature, such as toxicological reports on file with the USEPA as part of the Toxic Substance Control Act Section 8D, may become available from stakeholders and other interested parties during the external and stakeholder review process.

3.4 Determining the Critical Study and Endpoint

Development of an IDLH value begins with the critical evaluation and array of the available animal toxicity and human health effects data. In order to effectively evaluate the data, it is useful to array the following information:

Table 3–1. Literature search sources

Database	Link
Centers for Disease Control and Prevention (CDC)/ Agency for Toxic Substance and Disease Registry (ATSDR) ToxProfiles	http://www.atsdr.cdc.gov/toxprofiles/index.asp
ChemIDplus	http://chem.sis.nlm.nih.gov/chemidplus/chemidlite.jsp
EU, European INventory of Existing Commercial chemical Substances (EINECS)	http://ihcp.jrc.ec.europa.eu/our_labs/predictive_toxicology/information-sources/ec_inventory
EMBASE	http://www.embase.com/
National Library of Medicine (NLM), Haz-Map	http://hazmap.nlm.nih.gov/
NLM, Hazardous Substance Data Bank (HSDB)	http://toxnet.nlm.nih.gov/cgi-bin/sis/htmlgen?HSDB
International Agency for Research on Cancer (IARC)	http://www.iarc.fr/
Environmental Protection Agency (EPA) Toxic Substance Control Act (TSCA) Section 8(e) Notices	http://www.epa.gov/oppt/tsca8e/index.html
World Health Organization (WHO)/IPCS International Chemical Safety Card (ICSC)	http://www.ilo.org/public/english/protection/safework/cis/products/icsc/dtasht/index.htm
International Toxicity Estimates for Risk (ITER)	http://toxnet.nlm.nih.gov/cgi-bin/sis/htmlgen?iter
New Jersey Hazardous Substance Fact Sheets (NJ-HSFS)	http://web.doh.state.nj.us/rtkhsfs/indexfs.aspx
NIOSHTIC2	http://www2a.cdc.gov/nioshtic-2/default.asp
NLM, PUBMED	http://www.ncbi.nlm.nih.gov/pubmed/
National Institute for Occupational Safety and Health (NIOSH)/Registry of Toxic Effects of Chemical Substances (RTECS)	http://www.cdc.gov/niosh/rtecs/
NLM, Toxicology Literature Online (TOXLINE)	http://toxnet.nlm.nih.gov/cgi-bin/sis/htmlgen?TOXLINE
Web of Science	http://thomsonreuters.com/products_services/science/science_products/scholarly_research_analysis/research_discovery/web_of_science

- Description of the test species
- Health endpoints evaluated
- Exposure concentrations
- Critical effect levels (e.g., NOAELs, LOAELs, LC_{50} values)
- Duration of the exposure for the study.

Once this information is compiled, critical effect levels are adjusted to a 30-minute-equivalent concentration to derive a POD estimate for each

Table 3–2. Literature search key words

Search terms	
Acute	Irritation
Inhalation	Respiratory
Lethal	RD
Lethal concentration	Threshold
LC	Case study
Fatal	Poisoning
Fatality	Chemical identifiers

study or study endpoint. Through the application of the weight-of-evidence approach described in this document, the critical study that contributes most significantly to the qualitative and quantitative assessment of risk is selected as the basis of an IDLH value. Appendix A provides an example of how such information is compiled and used in the derivation of the IDLH value for chlorine (CAS# 7782-50-5). The weight given to each study in selection of a final POD is based on the reliability of the reported findings (as determined from an assessment of study quality), the relevance of the study type for predicting human effects from acute inhalation exposure, and the estimated 30-minute adjusted effect level.

3.4.1 Study Quality Considerations

For toxicology studies, quality considerations that affect the reliability of each study include the key elements of the study design and the adequacy of study documentation. For example, such aspects of study quality might include the following:

- Relevance of the exposure regimen to a single 30-minute inhalation exposure
- Quality of atmosphere generation system and analytical techniques used to assess exposure conditions
- Degree of evaluation of toxic endpoints
- Number of animals used and relevance of the test species to humans.

Other considerations for evaluation of study quality include the reliability of the cited data source, whether the study adhered to or was equivalent to current standards of practice (e.g., USEPA or Organisation for Economic Co-operation and Development [OECD] test guidelines), and whether good laboratory practices (GLPs) were followed. These considerations are evaluated for each study according to the general concepts outlined by Klimish et al. [1997]. Although a single authoritative guide to such study quality evaluation for epidemiology studies is not available, human effects data studies are judged on the basis of current standards of practice for conducting epidemiology or clinical studies [USEPA 1994; Federal Focus Inc. 1995; Lewandowski and Rhomberg 2005]. Consistency of effects across studies and consistency based on other information available about the chemical (e.g., oral data, structure-activity data) are used to assess the quality of individual studies.

Selection of the critical study to serve as the basis for the IDLH relies heavily on study quality considerations. A high-quality study might be chosen as the basis for the IDLH value, even if a lower IDLH value could be generated from a low-quality study, where the evaluation of quality casts doubt on the reliability of the study results. An LC_{50} value derived from a USEPA or OECD guideline–compliant acute lethality study with robust atmosphere generation and measurement systems may be selected over a lower LC_{50} value from an older study that used a static exposure chamber system and reported only nominal air concentrations or that used a small number of animals or non-standard test species.

3.4.2 Study Relevance Considerations

The weight-of-evidence approach requires a critical evaluation of each study as to its relevance to the ultimate goal of the IDLH value derivation—to develop a scientifically-based estimate of the 30-minute human threshold concentration for severe, irreversible or escape impairing effects. The methodology for developing IDLH values follows a hierarchical approach based on the following preference for data:

- Acute human inhalation toxicity data
- Acute animal inhalation toxicity data
- Data for longer-term inhalation studies
- Inhalation data for analogous chemicals (i.e., toxicological surrogates)
- Acute animal oral toxicity data.

The IDLH methodology described in this CIB follows similar principles but is based more on an overall weight-of-evidence approach that considers study reliability, quality (as discussed in Section 3.4.1), relevance, and the magnitude of the observed effect levels. The evaluation of study relevance includes the type and severity of the effects observed, study duration, and route of exposure.

Other considerations that will be addressed during the selection of the key data include the following:

- Primary versus secondary sources.
- Peer-reviewed versus non-peer-reviewed studies.

The term *primary data* refers to information obtained directly from original studies and reports, whereas *secondary data* refers to information summarized within reviews and monographs. Primary data are given more weight within the derivation of IDLH values, whereas secondary data are used to provide background and supporting information. An exception to this may occur when critical primary data are unobtainable and an IDLH value cannot be derived without being based on data contained in a secondary source. In such cases, the IDLH value may be based on the information contained within the secondary data source. Some secondary sources provide greater value than other sources. For example, authoritative secondary sources might include robust toxicity profiles that have undergone extensive review, such as the ATSDR Toxicological Profiles or EPA IRIS Toxicological Reviews.

Peer reviewed data are generally preferred as the basis of an IDLH value, over data obtained from non-peer-reviewed sources. For this reason, peer-reviewed data take precedent over non-peer-reviewed data within the IDLH methodology. Exceptions are made when issues with the peer-reviewed data are identified or if non-peer-reviewed studies are determined to be of higher quality. Non-peer-reviewed data may take precedent over peer-reviewed data in circumstances such as these:

- Non-peer-reviewed studies used standardized or guideline-compliant protocols, but available peer-reviewed studies used non-standardized protocols.
- The toxic effects reported in non-peer-reviewed studies align better with the health endpoints of interest (e.g., escape-impairing effects, irreversible effects, or lethality) than do the effects reported in peer-reviewed studies (e.g., mild irritation).
- Non-peer-reviewed studies demonstrate better biological and statistical significance due to increased sample size, selection of test species, or overall study design, in comparison with peer-reviewed studies.

Ultimately, the basis of an IDLH value will result from the weight-of-evidence approach incorporated into the CIB that reflects the relative strengths and weaknesses of all the data.

3.4.2.1 Relevance of the Type and Severity of the Effect

3.4.2.1.1 General considerations in identifying the severity of effects for IDLH derivation

Relevance of the effect is evaluated in the context of the goal for deriving an IDLH value (i.e., to develop a high-confidence estimate of the 30-minute human threshold concentration for severe, irreversible, or escape-impairing effects). Studies that identify with good precision the actual threshold for such effects are rare; therefore, usually it is necessary either to extrapolate from an effect level that is above a threshold, by relying on a lowest observed adverse effect level (LOAEL) for severe or escape-impairing effects, or to use a lower-bound estimate of the threshold by relying on a no observed adverse effect level (NOAEL) for severe or escape-impairing effects. In some cases, concentration modeling can be used to further refine such estimates on the basis of actual study concentrations. All of the data for effects relevant to the IDLH are evaluated and used in this effort, including data on mortality, severe or irreversible effects, and escape-impairing effects. Data on exposure levels causing less severe effects, which are below the threshold of interest, are useful as estimates of the NOAEL for severe effects or escape-impairment. Together, these data can describe the exposure–response relationship for the chemical of interest, which compares the estimated exposure concentration to the reported effects. Having an understanding of this relationship allows the potential region of the threshold concentration to be more accurately determined for the most-sensitive severe or escape-impairing effects.

Table 3–3 illustrates how the severity of effect is taken into account in determining the POD and IDLH value. In this case, human data are available for a 30-minute exposure that describe the concentration response, from no effects at 10 ppm to mild irritation at 20 ppm and severe irritation that was considered escape-impairing at 30 ppm. Thus, the threshold for an escape-impairing effect in humans is between 20 and 30 ppm for a 30-minute exposure, and the POD for the IDLH value would be 20 ppm. In this case, no concentration–response modeling was available to estimate the threshold for severe lacrimation and coughing. Application of a typical UF of 3 (see Chapter 4) to the NOAEL concentration of 20 ppm for mild irritation and coughing would generate an IDLH value of 7 ppm, which would be lower than appropriate on the basis of the absence of any irritant effects at 10 ppm. Thus, in this case, since the severity of the effects at 20 ppm was not considered escape-impairing, the appropriate IDLH value would be approximately 20 ppm or less on the basis of balancing consideration of the human effect level and variability in human sensitivity.

3.4.2.1.2 Consideration of lethality data

Datasets for acute toxicity are often limited to studies reporting mortality experience in acute animal toxicology studies or from case reports describing accidental human exposures that include estimates of the lethal concentration. Lethality data from acute toxicology studies in animals are commonly available, and many IDLH values are derived from such data. In such cases, information on the threshold for lethality is the preferred basis for an IDLH value, rather than an estimate of median lethal concentration (i.e., the LC_{50}). Lethality thresholds can be estimated from LC_{LO} values (the lowest concentration in the study that caused lethality) if the mortality incidence is relatively low (i.e., 10% or less) or can be based on concentration-response models. These models can be used to indicate the estimated response incidence (percent response) and whether the estimate is the maximum likelihood estimate or a lower confidence limit. For example, a commonly reported model value such as an LC_{01} value (the statistically derived air concentration that caused lethality in 1% of test animals) is the model estimated maximum likelihood concentration associated with an increased mortality incidence of 1% over control values. More recently, studies report lethality estimates with use of software that provides lower confidence estimates of the concentrations. For example, the USEPA provides free software for this purpose (available at http://www.epa.gov/ncea/bmds/). The output from the USEPA software is commonly reported as the benchmark concentration (BMC) for the maximum likelihood

Table 3–3. Consideration of severity of effect

Species	Endpoint—effect level, ppm	Duration, minutes	Comments
Human	NOAEL—10	30	No irritation
Human	NOAEL—20	30	Mild irritation and coughing
Human	LOAEL—30	30	Severe lacrimation and coughing

Abbreviations: LOAEL = lowest observed adverse effect level; NOAEL = no observed adverse effect level; ppm = parts per million.

estimate or the 95% lower-bound confidence limit on the BMC (BMCL). Thus, a $BMCL_{05}$ is the estimated 95% lower confidence bound on the concentration associated with a 5% increased lethality response above controls.

Such model-calculated values are preferred over LC_{LO} values, because they are not dependent on the actual concentrations tested and reflect the response at each concentration. Use of a lower confidence limit (i.e., the BMCL) also has the advantage of taking into account the uncertainty in the data and statistical power of the study. Frequently, the $BMCL_{05}$ (i.e., the lower 95% confidence limit on the concentration associated with a 5% response) and BMC_{01} (i.e., the central tendency estimate of the concentration associated with a 1% response) are both calculated for lethality data, and the lower value is used as the lethality threshold. The lower value is often the $BMCL_{05}$, due to the relatively wide confidence limits associated with the small sample size. An extensive discussion on the application of benchmark dose (BMD) within the development of acute emergency response guidelines has been included in the AEGL SOP [NAS 2001]. This includes key considerations, shortcomings, and uncertainty within the process. An alternative approach used for estimating a non-lethal exposure level from LC_{50} values has been applied in the AEGL methodology [NAS 2001; Rusch et al., 2009]. This approach uses 1/3 of the LC_{50} value as the POD to estimate the boundary between the lethality threshold and a non-lethal exposure level. When compared to LC_{01} values and $BMCL_{05}$ values for selected chemicals, in general, 1/3 of the LC_{50} value resulted in lower estimates of a non-lethal threshold [Rusch et al. 2009]; thus, this is a health-protective approach. Although the use of lethality data as the basis of an IDLH value is not ideal, the absence of concentration-response data may require the use of LC_{50} values as a POD.

Although estimates of a lethality threshold are preferred over other measures of lethal concentrations, in many cases, the only available data from acute lethality studies are LC_{50} values (i.e., concentrations associated with a 50% mortality incidence).[†] If LC_{50} value estimates are available for multiple species, then the lowest reliable LC_{50} value in the most relevant animal species is used for extrapolation to predict human response. If no data are available that favor the use of one animal species over another, then the most sensitive species is used after considering study quality. Multiple LC_{50} values may also be available from a single study, including values for both sexes individually and for the two sexes combined. In such cases, the data are evaluated for any clear difference between the sexes. If a clear difference exists, the LC_{50} from the more sensitive sex is used. If there is no clear difference, the combined LC_{50} value is used, since the combined data provide a higher statistical power.

Table 3–4 illustrates different lethality data that may be available. In the example cited, three different measures of lethality are available from the rat study: the LC_{50}, LC_{LO}, and $BMCL_{05}$. The selected POD for deriving the IDLH value would be the rat $BMCL_{05}$, because this value represents a defined response near the threshold for lethality and the data show that the rat is more sensitive than the mouse. In this case, the $BMCL_{05}$ resulted in the lowest derived value, but the $BMCL_{05}$ would generally be preferred, even if it was somewhat higher than the LC_{LO}, due to statistical variability related to the LC_{LO} and because the $BMCL_{05}$ reflects the variability in the data. The derived IDLH values reflect the application of UFs, addressing how far the data and endpoints are from the endpoint of interest. Because the goal is to estimate the threshold for the severe responses, a larger UF is applied to the LC_{50} value than is applied to measures around

---

[†] LC_{50} and BMC values are conceptually similar, although the BMC approach is a more recent innovation. Both values are determined by fitting a flexible mathematical curve to the data, and determining the concentration corresponding to a specified response. While various mathematical models can be fit to the data, the probit model is frequently used, as a flexible model that usually fits acute data well, particularly for lethality data (e.g., Fowles et al. [1999]).

Table 3–4. Consideration of lethality effects

Species	Endpoint/effect level (ppm)	Duration (minutes)	UF	Derived IDLH value (ppm)	Comments
Rat	$LC_{50}/1000$	30	30	33	Males and females combined
Rat	$LC_{LO}/400$	30	10	40	1/10 died
Rat	$BMCL_{05}/240$	30	10	24	Modeling done by the authors
Mouse	$LC_{50}/2000$	30	30	66	Males only

Abbreviations: $BMCL_{05}$ = lower 95% confidence limit on the concentration associated with a 5% response; IDLH = immediately dangerous to life or health; LC_{50} value = median lethal concentration; ppm = parts per million; UF = uncertainty factor.

the threshold for lethality, such as the $BMCL_{05}$ (see Chapter 4 for additional discussion of UF).

3.4.2.1.3 Consideration of escape-impairing effects

For effects other than mortality, reported health effects in both human and animal studies are classified as severe, irreversible, or escape-impairing. Identifying which effects may be escape-impairing is complicated by the fact that observed signs and symptoms in animals may differ from those expected to occur in humans. For example, the same underlying MOA that manifests as changes in respiration rate, nasal discharge, or altered activity level in an acute toxicity test in animals may be reported as intolerable irritation in humans. For this reason, guidance was developed that allows for more consistent assigning of comparative severity of observed effects (i.e., severe and irreversible versus non-severe; escape-impairing versus non-escape-impairing) for commonly observed adverse effects used as the basis of IDLH values. Appendix C provides the guidelines for classifying effects commonly seen in acute animal studies.

Generally, basing IDLH values on effects that can impair escape relates to consideration of irritation responses (e.g., severe eye burning or coughing) or impacts on the nervous system (e.g., headache, dizziness, drowsiness), although other effects (e.g., cardiovascular or gastrointestinal tract effects) may also be considered, when warranted. To facilitate a consistent approach, qualitative descriptions of severity have been developed with study results assigned to one of three categories: mild, moderate, or severe. The severity and the type of the effect are considered in determining whether escape impairment is likely. For example, moderate to severe eye irritation, but not mild irritation, is generally considered an appropriate basis for an IDLH value based on escape impairment. For effects on the CNS, narcosis or moderate dizziness is considered sufficiently adverse to impair escape, whereas effects such as headache are generally not considered as an adequate basis for the IDLH value unless described in the study as debilitating or occurring with other symptoms that directly impaired vision or mobility.

Additional consideration is needed for screening assays, such as the respiratory depression 50% (RD_{50}) assay and cardiac sensitization tests. The RD_{50} assay is a sensitive measure of sensory irritation, which occurs due to stimulation of trigeminal nerve endings in the cornea and nasal mucosa. These effects frequently are due to a decrease in respiratory frequency that occurs in some laboratory animals when exposed to chemical irritants. The RD_{50} value is considered as part of the overall weight of evidence and can be used to support the selection of a POD from other studies that identified the concentration that caused clinical signs of irritation or generated histopathologic changes consistent with moderate or severe irritant effects [Alarie 1981; ASTM 1984; Schaper 1993; Nielsen et al. 2007]. The RD_{50} value can also be used as the POD if no reliable LOAEL is available. However, the LOAEL is preferred over the RD_{50} value as a POD because of uncertainties in relating the respiratory depression response in rodents to potential clinical or tissue changes in humans that would be correlated with severe irritation in humans [Bos et al. 1992, 2002].

Cardiac sensitization is another sensitive endpoint [Brock et al. 2003; ECETOC 2009] that serves as the basis of some IDLH values. This endpoint reflects a serious effect in humans, which is characterized by the sensitization of the heart to arrhythmias. Cardiac sensitization can occur from exposure to some hydrocarbons and hydrocarbon derivatives which make the mammalian heart abnormally sensitive to epinephrine. This can result in ventricular arrhythmias and, in some cases, can lead to sudden death [Reinhardt et al. 1971]. The arrhythmia results from the hydrocarbon potentiating the effect of endogenous epinephrine (adrenalin), rather than a direct effect of exposure to the hydrocarbon. As described by NAS [2002], "the mechanism of action of cardiac sensitization is not completely understood but appears to involve a disturbance in the normal conduction of the electrical impulse through the heart, probably by producing a local disturbance in the electrical potential across cell membranes."

Cardiac sensitization is determined by injecting the test animal (usually dogs, but rodents are also used) with epinephrine to establish a background (control) response, followed by an injection of epinephrine during exposure to the chemical of interest. Different doses of epinephrine are often tested for the initial injection, and the dose of epinephrine chosen is the maximum dose that does not cause a serious arrhythmia [NAS 1996]. The test is very conservative, because the levels of epinephrine administered result in blood concentrations approximately 10 times the blood concentrations that would be achieved endogenously in dogs [Chengelis 1997] or humans [NAS 1996], even under highly stressful situations. Thus, even though scenarios where IDLH values would apply would be highly stressful, the cardiac sensitization test is considered a sensitive measure of a severe effect. Cardiac sensitization is relevant to humans, but because of the conditions of the assay, which focuses on the measurement of the response to a challenge injection with epinephrine, the assay itself is very sensitive [Brock et al. 2003; ECETOC 2009]. The sensitivity of the assay is considered in the weight-of-evidence approach when selecting the POD and in the selection of the UF.

3.4.2.1.4 Consideration of severe and irreversible effects

A variety of health effects may result from acute exposures that do not immediately impair escape (although over an extended time period these effects may be lethal). Severe adverse effects that are not immediately escape-impairing are evaluated on a case-by-case basis, by weighing considerations such as the need for medical treatment, the potential for altered function or disability, the potential for long-term deficits in function, and the likelihood for secondary symptoms that would be escape-impairing. These include severe, but reversible, acute effects such as hemolysis, chemical asphyxia, delayed pulmonary edema, or significant acute organ damage (e.g., hepatitis, decreased kidney function). If a chemical is suspected of generating such effects, then it is important to evaluate the design of the study to ensure that adequate time was allowed, following completion of the exposure period, to determine whether such latent effects of interest were assessed.

Irreversible target organ effects (e.g., permanent functional respiratory impairment or permanent neurological impairment) are also considered a sufficient basis for an IDLH value. As discussed further in the following paragraphs, data on irreversible effects of special interest (e.g., reproductive and developmental toxicity) or effects that have significant latency (e.g., cancer) are generally considered as an adequate basis for the IDLH value only when single-exposure studies have been conducted that evaluated these endpoints. For example, if reproductive or developmental studies involving short-term exposures (i.e., 1 day or less) are available and have adequately long observation periods to observe delayed effects, then they are considered in the development of the IDLH value; such studies can be informative regarding the potential for irreversible reproductive or developmental effects. These effects are considered in the overall weight-of-evidence analysis to ensure that the derived IDLH value is sufficiently protective against the most sensitive health endpoint, as described in the following paragraphs.

Standard developmental toxicity studies are not used directly because they typically involve repeated exposures (e.g., during all of gestation or from implantation through one day prior to expected parturition), and extrapolation from studies that involve long exposure periods thereby resulting in an unacceptable level of uncertainty. However, it is also recognized that some developmental effects can result from exposure during a critical window of development, and that the time in which the exposure is administered may be more important than exposure duration. Therefore, data from developmental studies are evaluated in the context of the overall weight-of-evidence analysis. For example, if developmental effects are seen, the data on MOA and the relative concentration response for maternal toxicity and fetal toxicity are evaluated to determine whether an increased UF is needed. Conversely, a potential IDLH value derived from systemic toxicity in the pregnant female can provide a health-protective, lower-bound estimate for the IDLH value, because the exposure duration of repeated days is much longer than the duration of interest—a single 30-minute exposure. Use of repeated-exposure studies in this manner can provide perspective to potential IDLH values derived from very high concentration acute studies where a large UF leads to relatively low IDLH values that are more than adequately protective. Information relating to key issues in the use of developmental toxicity data during the assessment of the health risks of acute exposure scenarios has been published [van Raaij et al. 2003; Davis et al. 2009]. These publications provide supplemental resources that will be used to refine the derivation of IDLH values based on developmental toxicity data.

Table 3–5 shows how developmental toxicity data can be used to help evaluate an appropriate lower-bound estimate for the IDLH value. In this case, the IDLH value is derived from the 60-minute LC_{50} value, as the lowest acute lethality value from the studies of relevant duration. (See Section 3.5 and Chapter 4.0, respectively, for discussion of the adjustment for durations other than 30 minutes and UF used to calculate the derived value.) A developmental toxicity study is also available, in which exposure was for 6 hours/day on gestation days 6 to 20. Because the developmental effect of decreased fetal body weight may have resulted from a single exposure during a critical window, the exposure duration is listed as 6 hours. Because this is a very health-protective assumption, the developmental toxicity study is not used as the basis for the IDLH value, since confidence in the actual acute exposure effect level is highly uncertain. However, the derived IDLH value does provide a lower-bound estimate, since we would not expect the LC_{50}-based IDLH value to be lower than the derived value from a repeat-exposure study for non-lethal effects. The IDLH derived from the LC_{50} is somewhat higher than from the repeated-exposure developmental toxicity study; thus, the overall findings are consistent with expectations and the overall dataset provides reasonable confidence in the selected value.

Like developmental toxicity studies, reproductive toxicity studies tend to involve repeated exposures and therefore usually are not used as the basis for an IDLH value. However, single-exposure reproductive toxicity studies that report irreversible or slowly reversible effects are considered in the development of IDLH values. In addition, findings of reproductive toxicity coupled with MOA data (e.g., data suggesting an effect on hormonal control) may suggest the use of an increased UF, if the available acute toxicity data are insufficient to evaluate the concentration–duration response for such effects.

Table 3–5. Consideration of developmental toxicity data

Species	Endpoint/effect level (ppm)	Duration (minutes)	Adjusted to 30 minutes (ppm)	UF	Derived value (ppm)	Comments
Rat	LC_{50}/1800	60	2268	30	75.6	2/4 died—not a calculated value
Rat	LOAEL/200	360	458	10	45.8	1/21 dams died; fetal weight decreased; significant reabsorptions at 300 ppm; 6 hours/day on days 6–20 of gestation

Abbreviations: LC_{50} = median lethal concentration; LOAEL = lowest observed adverse effect level; ppm = parts per million; UF = uncertainty factor.

As noted above, acute animal toxicity studies rarely include sufficient post-exposure monitoring to be useful for cancer assessment. Even when a study is sufficient for evaluating carcinogenicity following a single exposure (e.g., Hehir et al. [1981]), such as following vinyl chloride exposure, the data are usually insufficient for a quantitative calculation of cancer risk. Therefore, concern for carcinogenicity is addressed by consideration of adding a supplemental UF (see Chapter 4). The cancer risk at the potential IDLH value can also be estimated and compared with a chosen risk level (i.e., a 1 in 1,000 excess cancer risk) [NAS 2001]. The concentration corresponding to a specified risk level is not usually used as the basis for the IDLH value, because of the considerable uncertainty in extrapolating from a chronic study to a single exposure. However, if the estimated cancer risk at the IDLH value without the supplemental UF is below 1 in 1,000, then the supplemental UF is not used.

3.4.2.2 Relevance of the Exposure Duration for Acute Studies

Acute animal inhalation studies reviewed for the derivation of the IDLH value may use treatment regimens ranging from an exposure duration as short as a few minutes (e.g., <10 minutes) to several hours (e.g., 8 hours or more). Because the intended use of the IDLH value is for the prevention of adverse effects that may occur as a result of a single exposure for 30 minutes, the derivation of an IDLH value is ideally based on:

- Studies involving exposure for 30 minutes
- Studies that have information on the threshold for rapidly occurring escape-impairing effects
- Studies that include a sufficient observation period for potential severe delayed effects.

Acute studies of durations other than 30 minutes that provide information on escape-impairing effects and severe adverse effects are also desirable and used. Although inhalation studies of durations other than 30 minutes introduce uncertainties in extrapolating effects to a 30-minute duration, they are still used after being adjusted to a 30-minute-equivalent exposure duration, as discussed in detail in Section 3.5 on Time Scaling.

It is recognized that the ideal dataset applied during the derivation of an IDLH value will consist of high-quality 30-minute inhalation studies with effects in the severity range of interest. In most cases, such datasets are unavailable. Thus, when selecting among less-than-optimal study designs to identify the most appropriate critical study and POD, a weight-of-evidence approach is used to select the critical study. For example, within a given category of studies (e.g., acute lethality studies), preference is given to high-quality studies of the duration of interest (30 minutes) or involving minimal duration extrapolation (e.g., 20-minute exposure duration is preferred over a 4-hour exposure duration). However, the relative merits of a well-done study of longer duration versus a poorly conducted 30-minute study must be considered. A well-documented weight-of-evidence decision is even more important when there are no adequate acute inhalation studies in humans or animals. In such cases, consideration of all other available data is needed, including MOA information, repeated-exposure studies, studies of exposure routes other than inhalation (e.g., oral or direct-injection dosing), and studies with other (usually structurally related) chemicals. MOA understanding is particularly important in such situations and can determine such issues as whether route-to-route extrapolation is appropriate, the impact of using data from repeated-exposure studies, and which structurally related chemicals are appropriate to use by analogy. The following examples illustrate the impact of MOA on extrapolation decisions.

1. **For route-to-route extrapolation**: It is inappropriate to conduct route-to-route extrapolation for irritants, because they target the route of entry.

2. **For duration extrapolation**: It may be appropriate to extrapolate from repeated-exposure studies for irritants, since concentration is often a more important determinant of irritation than exposure duration. Irritation effects observed on the first day of exposure during a repeated-exposure study may be used as the basis of an IDLH value.

Repeated-exposure studies that identify subchronic or chronic systemic toxicity (rather than rapid-onset clinical signs) are not used quantitatively as the basis for deriving the IDLH value. However, considerations of these other toxicity metrics are included in overall database evaluation during the consideration of UF and to assess the reliability of estimates derived from acute studies. For example, if a well-conducted repeated-exposure study shows no adverse effect at a given concentration, then such a finding can help to determine the lower range of potential values for an IDLH value, since single acute exposures will usually identify a higher POD. In this way, repeated-exposure studies can provide a lower bound on the range of potential IDLH values for a chemical if the databases of acute studies are limited or of marginal quality.

Table 3–6 illustrates how scientific judgment is used in considering duration. In this example, only limited acute data are available for the chemical, including an RD_{50} study and one LC_{50}. However, some information on the effects of acute exposure can be extracted from clinical signs reported for a subchronic exposure study in which exposure was for 6 hours/day, 5 days/week, for 13 weeks. Clinical signs reported at 4.9 ppm were limited to eyes half-closed during exposure, an indication of eye irritation, but at a level that is not escape-impairing. However, at the next higher exposure level (15.3 ppm), the authors reported burning of the nose and eyes, as well as olfactory lesions. Although the lesions may have been related to the repeated exposure, it is reasonable to assume that the clinical signs of burning eyes and nose were observed during the first exposure, and that these effects would be escape-impairing. After consideration of time adjustments (see Section 3.5) and application of the appropriate UF (see Chapter 4.0), the LOAEL from the repeated-exposure study was used as the basis for the IDLH value, supported by the RD_{50}. A slightly higher IDLH value would have been calculated from the LC_{50}, but that value was not used, since it involves more extrapolation due to the severity of the response (lethality). Direct observations from the initial exposure during the repeated-exposure study were considered more reliable than using the RD_{50} value directly, based on the uncertainties in interpreting the RD_{50} assay.

3.4.2.3 Relevance of the Exposure Measurements

Animal inhalation studies are typically conducted using either whole-body or nose-only exposure. Both methods have strengths and limitations.

Whole-body exposure more closely simulates the situation for occupational exposure and includes the potential for exposure both via inhalation and via dermal exposure to the chemical in the air. However, in rodent studies, whole-body exposure may also involve ingestion exposure that is not relevant to humans, due to grooming of fur on which the chemical has deposited. Nose-only exposure avoids the potential for ingestion exposure, but also eliminates the potential for human-relevant dermal exposure, and may place the animals under additional stress, because of their being restrained during exposure. There is no default preference for one exposure scenario over the other. Instead, the studies and results should be examined to determine whether the limitations of either method preclude the use of certain studies. For example, the observation of overt gastrointestinal (GI) effects from whole-body exposure suggests the potential for confounding by ingestion. In general, both nose-only and whole-body exposures are considered together in the overall weight-of-evidence evaluation.

Well-conducted inhalation studies generally report both nominal concentrations (the concentration expected on the basis of the amount of chemical introduced into the exposure system) and the analytical concentration (the amount actually measured). The two values should be similar; if they are markedly different, the reasons and implications for the difference should be determined. Large differences may reflect difficulty in maintaining the exposure atmosphere (e.g., the chemical may be adhering to the exposure chamber walls) or other issues, and may indicate uncertain study quality. Larger differences between nominal and analytical concentrations may be seen with static exposure studies (where the chemical is introduced into the chamber at the beginning of the experiment), as opposed to dynamic studies (where the chemical is continuously circulated and the chemical concentration is actively maintained at the target level). Because the analytical concentration reflects the actual concentration to which the animals were exposed, the analytical concentration is usually used in IDLH value calculations. However, in some cases, the nominal concentration may more appropriately reflect the exposure conditions. For example, substances, such as trichloromethylsilane (CAS# 75-79-6), sulfur trioxide (CAS# 7446-11-9), uranium hexafluoride (CAS# 7783-81-5), and acetone cyanohydrin (CAS# 75-86-5), react with the moisture in air to produce a variety of hydrolysis products. Table 3–7 provides examples of hydrolysis products associated with the previously listed substances. Because the observed toxicity is due to both the parent chemical and the hydrolysis products, nominal concentration is a better indicator of toxicity, since it reflects the total burden of toxic constituents, whereas analytical concentration would reflect only the concentration of the parent compound [NAS 2009]. In such cases, the decision of whether to use nominal or analytical concentrations depends on the approach that would be used for air monitoring and whether it would capture only the parent compound or the parent compound and its hydrolysis products.

Care should also be used in considering the exposure units. For example, it is appropriate to use ppm only for gases and vapors because ppm in air refers to molecules of the chemical in air (rather than being on a weight basis). The units of mg/m^3 can be used for particulates and aerosols, as well as gases and vapors. Although exposures to gases and vapors are usually reported in ppm, care is needed to ensure that units are not confused. Units of ppm can be converted to mg/m^3 using the ideal gas law. At 1 atmosphere of pressure and room temperature (25° C), the conversion is as follows:

$$mg/m^3 = ppm \times molecular\ weight/24.45$$

Difficulties in the determination of exposure concentrations may arise because, at high concentrations, some vapors may condense into liquid droplets, resulting in exposures to a mixture of vapor and aerosol. Under such conditions, it is generally reasonable to assume that toxicity is due to the total mass of the chemical. However, it should be recognized that vapors and aerosols (e.g., solid particles and liquid droplets) are deposited differently in the respiratory tract on the basis of many factors, including the physiochemical properties of the chemical [USEPA 1994]. For this reason, the

Table 3–6. Use of scientific judgment

Species	Endpoint/effect level (ppm)	Duration (minutes)	Adjusted to 30 minutes (ppm)	UF	Derived IDLH value (ppm)	Comments
Mouse	RD_{50}/10.4	30	10.4	3	3.5	—
Rat	LC_{50}/125	240	250	30	8.3	—
Rat	NOAEL/4.9	360	11.2	3	3.7	6 hours/day, 5 days/week, 13 weeks; eyes half-closed during exposure
Rat	LOAEL/15.3	360	35.0	10	3.5	6 hours/day, 5 days/week, 13 weeks; olfactory lesions, burning nose and eyes

Abbreviations: IDLH = immediately dangerous to life or health; LC_{50} value = median lethal concentration; LOAEL = lowest observed adverse effect level; NOAEL = no observed adverse effect level; ppm = parts per million; RD_{50} value = median respiratory depression value; UF = uncertainty factor.

Table 3–7. Examples of hydrolysis products associated with selected chemicals

Chemical names	CAS#	Hydrolysis products	Health effects of hydrolysis products
Trichloromethylsilane	75-79-6	Hydrochloric acid (CAS# 7647-01-0)	Respiratory tract and eye irritation
Sulfur trioxide	7446-11-9	Sulfuric acid (CAS# 7664-93-9)	Respiratory tract and eye irritation
Acetone cyanohydrin	75-86-5	Hydrogen cyanide (CAS# 74-90-8); Acetone (CAS# 67-64-1)	Respiratory tract and eye irritation
Uranium hexafluoride	7783-81-5	Uranyl fluoride (CAS# 13536-84-010); Hydrogen fluoride (CAS# 7664-39-3)	Respiratory tract and eye irritation

Abbreviation: CAS # = chemical abstract service number

toxicity related to vapor exposure and aerosol exposure to the same concentration (e.g., mg/m^3) of a substance may be somewhat different if respiratory tract effects are of concern.

IDLH values derived for aerosols will reflect the relevant size fraction. Specific recommendations relating to size fractions for aerosols are included in the chemical-specific IDLH support documentation when sufficient data are available. The most appropriate size fraction is driven by the nature of acute toxicity observed. If such data are not available, the chemical-specific IDLH support documentation for the aerosol will note that the size fraction that represents the greatest hazard could not be determined. In such cases, total inhalable particulate is used as the basis for the IDLH value.

3.4.2.4 Other Issues of Study Relevance—Use of Surrogates and Route Extrapolation

When neither human nor animal acute inhalation data are sufficient to derive an IDLH value for a chemical of interest, other approaches are considered, depending on the understanding of the MOA and availability of data. Available information on surrogates, or related compounds, primary metabolites, or key breakdown products (e.g., secondary chemical products formed from hydrolysis due to moisture in the air) that are closely related to the chemical of interest can be used when inadequate information is available for the chemical of interest. As an example of the use of a related compound during the derivation of an IDLH value, bromine pentafluoride (CAS# 7789-30-2) and chlorine pentafluoride (CAS# 13637-63-3) differ only in the primary halogen atom. Because of their similarities, bromine pentafluoride can be used as a surrogate for chlorine pentafluoride, and the limited toxicity data available for bromine pentafluoride indicate that its toxicity is comparable to or slightly less than that of the chlorine compound. Another example is the assessment of the acute inhalation hazard of an entire chemical class on the basis of the data for a single compound; the NAS/NRC drafted AEGL values for multiple chlorosilanes and metal phosphides with use of this approach [NAS 2007, 2009]. This approach takes advantage of knowledge about the MOA and the actual form of the toxicity of related chemicals to use the entirety of the data for the class of chemicals to develop exposure values. For example, for the chlorosilanes the primary cause of the acute effect of interest (irritation) is hydrolysis in moist air to form hydrochloric acid. Thus, for the series of related chlorosilanes, the IDLH value can be derived from actual testing data for the most data-rich member of the family and by adjusting the IDLH value for other members according to the respective amounts of chlorine atoms produced during hydrolysis. A refinement of the use of surrogate chemicals or information on classes of related chemicals is to use data on the relative potency, when adequate data are available to quantitatively compare the chemical of interest with the surrogate but data for the chemical itself are not sufficient to develop an IDLH value. In such cases, the toxicity threshold is much better understood for the surrogate than for the chemical of interest, but the threshold for the chemical of interest can be adjusted on the basis of relative potency.

When a surrogate or relative-potency approach is used, it is necessary to consider the uncertainties associated with using a limited database for the chemical of interest versus the uncertainties associated with extrapolation from a surrogate chemical. An example of extrapolation from a breakdown product is the chemical reaction that causes acetone cyanohydrin to form HCN and acetone. The acute toxicity of acetone cyanohydrin is driven by exposure to an equimolar (i.e., having an equal number of moles) equivalent to HCN. Thus, the acute toxicity data for HCN can serve as a surrogate and basis of an IDLH value for acetone cyanohydrin [NAS 2002, 2005]. Use of such surrogates is not necessary when adequate information on the primary chemical is available. In addition, if a surrogate is being considered as the basis for the IDLH value, it is important to consider whether other aspects of toxicity are associated with the parent chemical and whether these aspects are adequately addressed by the surrogate. For example, acetone cyanohydrin causes irritant effects that are not seen with exposure to HCN, but the most potent escape-impairing effects are secondary to cyanide action as a metabolic toxicant. This results in HCN being the most valid surrogate for acetone cyanohydrin.

If no adequate inhalation data are available for the chemical of interest or for a potential surrogate, an IDLH value may be derived by extrapolation from studies that used exposure routes other than inhalation, such as oral or intraperitoneal (i.p.) dosing studies. As noted above, this route-to-route extrapolation is appropriate only if the effect of interest is systemic (i.e., involves absorption into the systemic blood circulation for distribution to an internal target tissue). Route extrapolation (e.g., from oral or i.p. dosing studies) is not appropriate if the chemical's primary relevant effects for IDLH development are as an irritant, or if it is expected to target the route of entry (i.e., respiratory tract) as the most sensitive end point. The ideal approach is to use a physiologically based pharmacokinetic (PBPK) model to conduct the route-to-route extrapolation, but it is rare that such data would exist (particularly for a chemical for which the inhalation data are insufficient to directly derive an IDLH value). In the absence of such a PBPK model, the approach is to estimate the concentration to which a 70-kg worker could be exposed in order to receive the equivalent *systemic* dose to that delivered in the oral or i.p. study. The 30-minute concentration is estimated by multiplying the animal dose data by the worker body weight (to reach a systemic dose), and dividing by the volume of air inhaled per work day, as shown in this equation:

$$\text{Systemic dose equivalent [mg/10 m}^3\text{]} = \frac{\text{oral or i.p. dose [mg/kg]} \times 70 \text{ kg}}{1.5 \text{ m}^3}$$

This conversion is a health-protective estimate of the air concentration that would result in the systemic dose, since a worker breathing at a rate of 50 liters per minute (L/min) for 30 minutes would inhale 1.5 m³ of air. The basis for this decision is discussed in greater detail in Appendix E.

A second consideration in applying route-to-route extrapolation is the impact of first-pass metabolism. First-pass metabolism, also known as pre-systemic metabolism, refers to the metabolism of a chemical delivered from the GI tract directly to the liver via hepatic blood flow, before distribution to the general systemic circulation. First-pass metabolism by the liver generally decreases systemic exposure to the parent chemical following oral exposure when compared with inhalation exposure. More precisely, first-pass metabolism via the respiratory tract tends to be of smaller magnitude than for the liver resulting in increased systemic exposure to the parent chemical and decreased exposure to alternative organ systems to metabolites formed in the lungs. Quantitatively addressing the implications of first-pass metabolism is often difficult, and use of a surrogate for which inhalation data are available is considered to provide greater weight of evidence for chemicals where first-pass metabolism plays an important role. Comparing IDLH values derived from different approaches (e.g., using a surrogate versus using route-to-route extrapolation) can provide information on possible uncertainties involved and may help to set the range of reasonable IDLH values. Finally, since this approach is based on systemic dose, it assumes equal absorption via both routes (unless a separate correction is made) and ignores issues related to the physical characteristics of the chemical (e.g., gas/vapors versus particulate) and implications of particle size and dosimetry (i.e., determination of respiratory tract region deposition fractions). Where quantitative adjustments for differing routes of exposure are uncertain, this issue is further considered in the selection of additional UFs. Additional considerations for conducting route-to-route extrapolations are described in several guidance documents (e.g., [USEPA 1994; NAS 2001]).

3.5 Time Scaling

A critical consideration in developing IDLH values is accounting for exposure duration and the extrapolation from the experimental exposure duration to the duration of interest (i.e., 30 minutes).

The methods used for these extrapolations in the development of IDLH values are similar in many ways to the Standing Operating Procedures (SOPs) outlined by the NAS for the development of AEGLs [NAS 2001]. Issues to be considered include evaluation of the chemical's MOA and how that is reflected in key drivers of toxicity (concentration vs. time); modifications to Haber's rule; and methods for calculating n in the ten Berge modification to Haber's rule. These issues are discussed briefly in the following paragraphs and in more detail in NAS [2001].

The toxicity of airborne chemicals depends on both exposure concentration and exposure duration, as well as physiochemical properties that affect respiratory deposition and systemic absorption. Ideally, information from validated PBPK or biologically based dose–response (BBDR) models is used for time extrapolation, but such information is rarely available. In the absence of such models, simpler concentration–time relationships are used. Historically, particularly for extended exposure durations, toxicity was described as the simple product of concentration (Conc) and time, so that Conc × time = k, a constant. In other words, if $Conc_1 \times time_1 = Conc_2 \times time_2$, then the toxicity would be the same. This relationship is described as Haber's law, or Haber's rule [Haber 1924].

The key assumption embedded in the relationship of Haber's rule is that damage (or depletion of protective tissue response) is irreversible and, therefore, that toxicity is cumulative, related to the total dose of the chemical [NAS 2001]. This assumption is generally not true for single acute exposures [NAS 2001]. For example, toxicity due to asphyxiants (e.g., argon or nitrogen) is related to the peak concentration of the chemical, rather than the cumulative dose. Sensory irritation and transient acute CNS effects may also be influenced more by the exposure concentration than the exposure duration.

Further investigation into the relationship between concentration, duration, and toxicity was conducted by ten Berge et al. [1986], who proposed the following relationship between Conc and duration (time, t): $Conc^n \times t = k$. These investigators examined the data on 20 irritant and systemically acting gases and vapors; the results of this investigation indicated that n was ≤3 for lethality data from 18 of the 20 chemicals. This study is one of the primary published sources for values of n. Furthermore, based on the finding in this study that an n of 3 covers 90% of the chemicals in the dataset, the default value of an n for extrapolating from longer durations to shorter durations was chosen to be 3, as a health-protective approach.

The following approach is used in extrapolating across durations within the IDLH methodology:

1. No extrapolation is needed if the study of interest involved exposure for 30 minutes; the empirical data are used directly.

2. If information on the value of n is available from the original paper of ten Berge et al. [1986] or from authoritative reviews (e.g., AEGL documents), then that value is used. Note, however, there are caveats to the use of the ten Berge data, and other considerations in the choice of n. In general, a published value of n will be used directly only for studies reporting the same effect or effects related to the same underlying toxic mode of action. Use of the published values of n for application to studies conducted in different species or for different effects is done on a case-by-case basis, with rationale provided.

3. If no value of n is available in the literature, n can be mathematically derived directly from the key studies of interest and applied with the same caveats as noted in item 2.

4. If the data are not available to support the derivation of n, then a default of 1 is used if the duration of the study of interest is less than 30 minutes, in which case the ten Berge equation defaults to Haber's rule. Conversely, if the duration of the study of interest is more than 30 minutes, then the default of 3 is used for n. This approach generally yields health-protective estimates for the 30-minute equivalent POD, as shown in Appendix E–2.

5. In limited cases, the overall dose–response data and the mode of action information may suggest that the observed acute effects are independent (or nearly independent) of exposure duration, and that exposure concentration can be used with no further duration adjustment. If the POD is used from studies of durations other than 30 minutes without adjusting to a 30-minute equivalent value via the ten Berge correction, the rationale for this decision will be described in the documentation of the IDLH value.

Additional information and illustration on the application of time scaling within the IDLH methodology are included in Appendix E–2.

3.6 Inclusion of Safety Considerations

Safety hazards are considered during the derivation of IDLH values to ensure the protection of worker safety and health. One particular consideration in the derivation of IDLH values is the potential for explosive concentrations of a flammable gas or vapor to be achieved at toxicologically relevant air concentrations. Maintaining safety considerations in the process for this methodology update is consistent with the prior method used to develop IDLH values. For gases and vapors, NIOSH has adopted a threshold of 10% of the LEL as a default basis for the IDLH values based on explosivity concerns. This threshold aligns with the airborne concentrations of a flammable gas, vapor or mist identified by OSHA as a hazardous explosive condition [29 CFR 1910.146(b)]. In such events, when the air concentration that corresponds with 10% of the LEL is less than the health-based value using the approach outlined in Chapter 3, this air concentration will become the default IDLH value. The following hazard statement will be included in the support documentation: "The health-based IDLH value is greater than 10% of the LEL (>10% LEL) of the chemical of interest in the air. Safety considerations related to the potential hazard of explosion must be taken into account." In addition, the notation (>10% LEL) will appear beside the IDLH value within the *NIOSH Pocket Guide to Chemical Hazards* [NIOSH 2005] and other NIOSH publications.

For dust, the equivalent default approach would be using 10% of the minimum explosive concentration (MEC). However, determining the combustibility of dusts is too complex to assign a single default measure. Dust combustibility and explosivity are dictated by the relationships among substance and scenario-specific factors including (1) particle size distribution, (2) minimum ignition energy, (3) moisture content, (4) explosion intensity and (5) dispersal in air [Cashdollar 2000]. The ability to quantify combustible dust specific concentrations for application of an IDLH is often not possible given the absence of critical chemical-specific data, such as the MEC or the other previously identified factors. NIOSH will critically assess the explosive nature of a dust when sufficient technical data are available. If determined to be appropriate, the findings of this assessment will be incorporated in the derivation process to ensure that the IDLH value protects against both health and safety hazards. When a dust has been identified as combustible, NIOSH will include the following hazard statement: "Dust may represent an explosive hazard. Safety considerations related to hazard of explosion must be taken into account." In addition, the notation (Combustible Dust) will appear beside the IDLH value in the *NIOSH Pocket Guide to Chemical Hazards* [NIOSH 2005] and in other NIOSH publications. Supplemental information on the combustibility of dust can be located on the OSHA Combustible Dust webpage (http://www.osha.gov/dsg/combustibledust/).

This page intentionally left blank.

4 Use of Uncertainty Factors

4.1 Application of Uncertainty Factors

As noted in prior sections of this CIB, the first step in the development of an IDLH value is to determine POD estimates adjusted to a 30-minute-equivalent exposure. However, in many cases the available POD values need to be further adjusted to develop an IDLH value that protects workers from potential lethal, severe or irreversible, or escape-impairing health effects. Thus the IDLH value can be represented as:

$$\text{IDLH Value} = \frac{\text{POD (e.g., 30-min-equivalent } LC_{50} \text{ value, } LC_{LO} \text{ value, LOAEL, or NOAEL)}}{\text{Total UF}}$$

The application of UFs is needed to account for uncertainties related to extrapolation from the concentration that caused effects in the selected toxicity study to those that would be expected to be below the threshold for such effects in workers exposed for up to 30 minutes. For example, if the most appropriate POD was an LC_{50} value in rats from a 30-minute exposure study, then use of this value directly as the IDLH value would clearly not be acceptable since a sub-threshold concentration for humans is needed. Dividing the selected POD, such as the LC_{50} value in this example, by an additional UF would then reduce the IDLH value to a lower concentration well below the LC_{50} value.

In general, the UFs need to address all key areas of uncertainty that result from extrapolating from the available studies. Most organizations that develop exposure values/limits consider the following key areas of uncertainty:

- **Interspecies variability in sensitivity:** This area addresses differences in sensitivity between the test species (e.g., mouse, rat, etc.) and the average human for the population of interest (i.e., in the context of IDLH application, workers).

- **Human variability in sensitivity:** This area addresses differences in sensitivity between the average human from the population of interest to the sensitive component of the population of interest.

- **Severity of effect:** Because the IDLH value is intended to be below a concentration that will cause death or severe, irreversible, or escape-impairing effects, the UF needs to account for extrapolation from a POD that caused such responses in the selected toxicology study to a concentration below the threshold for these effects.

- **Duration of exposure:** Some organizations that develop exposure values/limits include consideration of the duration of the study that served as the POD in the UF determination and its relevance to the duration of interest. In the context of IDLH development, this area of uncertainty is addressed through duration adjustments of the POD rather than the explicit application of a UF.

- **Other database deficiencies:** When datasets available to develop IDLH values are very limited, it is necessary to account for the possibility that the available studies did not identify the most sensitive endpoint relevant to IDLH development. In such cases it is appropriate to increase the UF to account for this uncertainty.

An approach used by many organizations, such as by USEPA for developing reference concentrations [USEPA 1994] and for the AEGL process [NAS 2001], involves consideration of these separate areas of uncertainty and the multiplication of UFs for each of these areas to derive the final cumulative UF.

The IDLH methodology is a modification of this approach that blends the rigor of full consideration of the relevant areas of uncertainty embedded in the USEPA and AEGL approaches with the flexibility to fully use the limited data from multiple lines of evidence often encountered in IDLH development. Overall, the assignment of UFs for IDLH derivation includes two steps:

1. Selection of an appropriate preliminary UF range
2. Modification of this preliminary range to select a final value.

The preliminary UF ranges are based on consideration of the study design and the adverse health effect occurring at the POD. Use of a preliminary range of values helps to ensure consistency in application of UFs within the IDLH development effort for diverse chemicals. However, modification of the UF is often required on the basis of unique issues arising from the review of the database for each unique chemical. Thus, the IDLH methodology captures the need to use a consistent approach for UF application while maximizing the ability to make informed decisions based on weight-of-evidence considerations.

4.2 The NIOSH IDLH Value Uncertainty Factor Approach

As discussed regarding the overall UF approach, the analysis focuses on the weight-of-evidence approach using all the relevant data. Thus, a range of preliminary UFs is shown for each of the typical types of effect levels that are available as a POD. However, the final UF applied is determined from the weight-of-evidence evaluation for each chemical that allows for modifying the preliminary UF on the basis of additional considerations unique to the dataset. The preliminary UF ranges are shown in Table 4–1. The most common UFs for a given data type are shown, but the range indicates how this value is commonly adjusted up or down according to the entirety of the database, as described further in this section. The preliminary UFs are applied as multiples of 1 or 10, with use of an intermediate value of 3. The value of 3 represents one half of the log10 unit (3.16 rounded to 3) as the minimum increments that are used for the UF adjustments to reflect the level of precision for such an approach. Although the value of 3 is used in place of 3.16 during the discussion of UFs, caution should be applied when multiplying UFs of 3 together. For

Table 4–1. Typical UF ranges

Point of departure	Typical UF range*
LC_{50} (in an animal study)	10 to 100
LC_{01}, LC_{Lo}, or $BMCL_{10}$ for lethality in animals	3 to 30
LC_{Lo} in humans	1 to 10
LOAEL for an escape-impairing or irreversible effect in animals	3 to 30
NOAEL for an escape-impairing or irreversible effect in animals, or animal RD_{50}	1 to 10
LOAEL for an escape-impairing or irreversible effect in humans	1 to 10
NOAEL for an escape-impairing or irreversible effect in humans	1 to 3

Abbreviations: $BMCL_{10}$ = lower confidence limit on the concentration associated with a 10% response; IDLH = immediately dangerous to life or health; LC_{01} = the statistically derived air concentration that caused lethality in 1% of test animals; LC_{50} = median lethal concentration; LC_{Lo} = lowest concentration of a substance in the air reported to cause death; LOAEL = lowest observed adverse effect level; NOAEL = no observed adverse effect level; UF = uncertainty factor.
*Typical UF Range is based on the information presented in Appendix D.

example, when multiplying two UFs of 3 together (e.g., 3 × 3), the product will be 10, not 9. This is also illustrated by the multiplication of three UFs of 3, that is, 3 × 3 × 3 will equal 30, not 27.

Selection of values other than the preliminary UF for deriving an IDLH value is common, reflecting the use of a weight-of-evidence approach and the sometimes-conflicting data from multiple lines of evidence. Common situations that lead to movement away from the preliminary UF value relate to evaluation of data for the areas of uncertainty and extrapolation noted in the prior section.

- **Interspecies variability in sensitivity:** If chemical-specific data are available to help determine the magnitude of the differences in species sensitivity, then such data are used to refine the size of the final UF. For example, if information about specific sensitivity due to differences in species metabolism is available, the UF applied to the POD from an animal study is adjusted accordingly (either up or down, depending on the data). If health effects data that serve as the POD are from human studies, then the UF would not need to address this area of uncertainty.

- **Human variability in sensitivity:** If chemical-specific data are available to help determine the magnitude of the variability in human sensitivity, then such data are used to refine the size of the final UF. If health effects data that serve as the POD are from a sensitive human group (e.g., non-smoking, young adult females for a clinical study of nasal irritation [Shusterman et al. 2003]), then the UF would be smaller in addressing this area of uncertainty. Because IDLH values are used in occupational applications, the range of variability that needs to be covered in applying the UF is expected to be less than for development of exposure values/limits meant to protect sensitive members of the general public. Conversely, if additional data do not include sensitive populations (e.g., asthmatics who may be exposed to respiratory irritants), then a larger UF may be selected.

- **Severity of effect:** The size of the adjustment needed would reflect the severity of effect observed at the POD. This is reflected in the preliminary UF ranges shown in Table 4–1. For example, as shown in the table, to derive an IDLH value that protects from severe effects, a larger margin would be needed between an LC_{50} value and the IDLH value than would be needed in between a $BMCL_{05}$ for an escape-impairing effect and the IDLH value. The range of preliminary values incorporates this consideration of effect severity.

  The consideration of the severity of effect also addresses the slope of the concentration–response curve. Steep concentration–response curves and high-quality data may result in UFs at the lower end of the range. Steep concentration–response curves represent estimates of responses that decrease rapidly with decreasing exposure concentrations, so that a smaller UF may be warranted to reach the response level in the concentration–response curve, compared with a more shallow concentration–response curve. Thus, if the concentration–response curve is very steep, a factor of 10 (rather than the preliminary UF of 30) may be applied to an LC_{50} value, based on consideration of the overall database. This is because there is less than a factor of 3 between the LC_{50} and the (actual or estimated) LC_{01} value.

- **Duration of exposure:** For most acute limits, including IDLH values, acute studies are typically used directly as the basis for the POD. Thus, the available studies are generally representative of the overall duration of interest (exposure for a single day or less). Further refinements to account for uncertainties in duration extrapolation, such as between a 4-hour study and the 30-minute duration of interest for IDLH development, are addressed in the time-scaling adjustment to the POD (see Section 3.5), rather than as a consideration for the UF value. However, significant uncertainties may need additional consideration if the available study is limited in design or outside the immediate duration range of interest. For example, if only repeat-exposure studies were available for a chemical to serve as the POD, and the observed effects were not clearly due only to initial acute exposures, then the use of such a POD might justify a smaller UF.

- **Other database deficiencies:** A UF at the higher end of the typical range (e.g., a UF of 10 instead of 3) is often used if major uncertainties or additional significant concerns are identified. If a database is very deficient, then the UF might be increased. This approach is often used if the only reliable data are lethality data from a single acute study. Other considerations for database deficiency relate to the potential for effects that were not evaluated in the available studies. For example, the higher end of the range may be used if the data indicate that the chemical is a sensory irritant and the data are insufficient to derive an IDLH value (e.g., due to inappropriate exposure durations) but indicate a large margin between concentrations causing severe irritation and those causing death. Other data gaps that may affect the size of the final UF reflect specific endpoints of concern. For example, a UF from the higher end of the range may be used if a chemical is a known or likely carcinogen or a developmental toxicant, with evidence that acute exposures may be of concern.

The examples in Appendix A highlight how these weight-of-evidence considerations are applied to select UFs and derive potential IDLH values.

4.3 Research Support for the NIOSH Uncertainty Factor Approach

The UF approach used for deriving IDLH values is based on a review of NIOSH research efforts, approaches used by other organizations that establish acute exposure limits/values, and other independent research.

The NIOSH approach is similar to that of other agencies in terms of the areas of uncertainty accounted for in determining the appropriate value of the final UF. Although the NIOSH approach does not assign an individual factor for each area of uncertainty, there is generally good agreement between the NIOSH UF and the UF embedded in derivation of AIHA ERPG values and the cumulative UF used for derivation of the AEGL values. As expected, there is not complete alignment between these values, because of differences in application of IDLH values versus other types of acute exposure limits. In particular, the UF applied to the IDLH value is often smaller than for deriving the ERPG or AEGL values, which results in a larger final exposure limit for IDLH values compared to these other guidelines. For example, differences often arise because of the explicit inclusion of potentially sensitive members of the general population (e.g., children, elderly, and individuals with health impairments) during the establishment of community-based acute exposure limits, such as the ERPG and AEGL. The IDLH values do not take into consideration the potentially sensitive members of the general population because it is assumed that they will not be substantially represented in the workforce for the purposes of considering average population responses. However, in some cases such populations may be considered when a chemical has specific effects on a target population that is well-represented in the expected worker population. An example would be an agent that has significant impacts on asthmatics. In such cases, health effects data from asthmatics that have been exposed to the agent would be appropriate for defining the POD as the basis for deriving an IDLH value.

To further verify that the preliminary ranges of the UF are supported by existing data, NIOSH conducted an analysis of acute toxicity data to determine the appropriate size of the UF for extrapolating from various points of departure to derive IDLH values that would be expected to protect from lethal, severe, irreversible, or escape-impairing effects in humans. Two approaches were used: one based on a detailed evaluation of acute toxicity data for 20 chemicals, and the second based on data for 94 chemicals taken from the documentation for IDLH values and consideration of MOA.

From these data compilations for chemicals with robust datasets, the ratios between animal lethality values commonly used as the POD for developing the IDLH value (e.g., LC_{50} values) and the effect level for lethality or other non-lethal effects in humans were determined for each chemical. The distribution of these ratios was analyzed, and the median value and 95th percentile value for each comparison were derived (see Appendix D). The resulting median values and upper-bound estimates for these case study chemicals were used to verify that the range of total UFs adopted in the IDLH methodology adequately accounts for the value that should be applied to an animal-based endpoint to protect from severe or escape-impairing effects in humans.

The analysis found that animal lethal concentrations and human effect thresholds (both LC_{LO} values and LOAELs for severe or escape-impairing effects) were generally correlated, such that chemicals with low animal LC_{50} values tended to have low human lethality thresholds and cause severe or escape-impairing effects in humans at low concentrations. This finding was important to support the approach of developing preliminary UF ranges that could be used to address protection from non-lethal effects when extrapolating from data from acute animal studies. Additional analyses were conducted by MOA category (e.g., irritant, CNS depressant, or "other") to determine if different UF ranges could be applied on the basis of a chemical's MOA. However, statistically significant differences were not found among the MOA categories. Thus, this further refinement to the approach for developing a preliminary UF to address effect severity by MOA category has not been applied for IDLH derivation. Overall, comparison of the median values to the UF ranges in Table 4–1 showed that the most common value is typically above or in the range of the median value for the comparison dataset. This result is also consistent with other evaluations that analyzed effect-level ratios from acute toxicity studies (e.g., Rusch et al. 2009). Additional results, as well as the results of the second approach, are presented in Appendix D.

This page intentionally left blank.

References

ACGIH [2009]. Threshold limit values (TLVs®) and biological exposure indices (BEIs®). Cincinnati, OH: American Conference of Governmental Industrial Hygienists.

AIHA [2006]. AIHA Emergency Response Planning (ERP) Committee procedures and responsibilities. Fairfax, VA: American Industrial Hygiene Association, http://www.aiha.org/get-involved/AIHAGuidelineFoundation/EmergencyResponsePlanningGuidelines/Documents/ERP-SOPs2006.pdf.

AIHA [2008]. Emergency response planning guidelines (ERPG) and workplace environmental exposure levels (WEEL) handbook. Fairfax, VA: American Industrial Hygiene Association Press.

Alarie Y [1981]. Dose-response analysis in animal studies: prediction of human responses. Environ Health Perspect 42:9–13.

ASTM [1984]. E981-84: standard test method for estimating sensory irritancy of airborne chemicals. Vol. 11.04. Philadelphia, PA: American Society for Testing and Materials.

Bos PM, Zwart G, Reuzel PG, Bragt P [1992]. Evaluation of the sensory irritation test for the assessment of occupational health risk. Crit Rev Toxicol 21:423–450.

Bos PM, Busschers M, Arts JH [2002]. Evaluation of the sensory irritation test (Alarie test) for the assessment of respiratory tract irritation. J Occup Environ Med 44:968–975.

Brock WJ, Rusch GM, Trochimowicz HJ [2003]. Cardiac sensitization: methodology and interpretation in risk assessment. Regul Toxicol Pharmacol 38(1):78–90.

Cal/EPA [2010]. Toxicity criteria database. Sacramento, CA: California Environmental Protection Agency, Office of Environmental Health Hazard Assessment, http://oehha.ca.gov/risk/ChemicalDB/index.asp.

Cashdollar KL [2000]. Overview of dust explosibility characteristics. J Loss Prev Process Ind 13:183–199.

Chengelis CP [1997]. Epinephrine sensitivity of the canine heart: a useful test. In Snyder R, Bakshi KS, Wagner BM. Abstracts of the Workshop on Toxicity of Alternatives to Chlorofluorocarbons. Inhal Toxicol 9:775–810.

Craig DK, Davis JS, Hansen DJ, Petrocchi AJ, Powell TJ, Euccinardi TE Jr [2000]. Derivation of temporary emergency exposure limits (TEELs). J Appl Toxicol 20:11–20.

Davis A, Gift JS, Woodall GM, Narotsky MG, Foureman GL [2009]. The role of development toxicity studies in acute exposure assessments: Analysis of single-day vs. multiple-day exposure regimens. Regul Toxicol Pharmacol 54(2):134–142.

ECETOC [2009]. Technical report no. 105: evaluation of cardiac sensitization test methods. Brussels, Belgium: European Centre for Ecotoxicology and Toxicology of Chemicals [ISSN-0773-8072-105].

Federal Focus Inc. [1995] Principles for evaluating epidemiologic data in regulatory risk assessment. Developed by an expert panel at a conference in London, England, October 1995. Washington, DC: Federal Focus Inc.

Fowles JR, Alexeeff GV, Dodge D [1999]. The use of benchmark dose methodology with acute inhalation lethality data. Regul Toxicol Pharmacol 29(3):262–278.

Haber F [1924]. Zur Geschichte des Gaskrieges. In: Fünf Vorträge aus den Jahren 1920–1923. Berlin: Springer-Verlag, pp. 76–92.

Hayes AW, ed [2008]. Principles and methods of toxicology. 5th ed. Boca Raton, FL: CRC Press.

Hehir RM, McNamara BP, McLaughlin J, Willigan DA, Bierbower G, Hardisty JF [1981]. Cancer induction following single and multiple exposure to a constant amount of vinyl chloride monomer. Environ Health Perspect 41:63–72.

IUPAC [2007]. International Union of Pure and Applied Chemistry glossary of terms used in toxicology. 2nd ed. Pure Appl Chem 79(7):1153–1344, http://sis.nlm.nih.gov/enviro/iupacglossary/frontmatter.html.

Klimish H-J et al. [1997]. A systematic approach for evaluating the quality of experimental toxicological and ecotoxicological data. Regul Toxicol Pharmacol 25:1–5.

Lewandowski TA, Rhomberg LR [2005]. A proposed methodology for selecting a trichloroethylene inhalation unit risk value for use in risk assessment. Regul Toxicol Pharmacol 41:39–54.

MDH [2010]. Health risk values (HRVs) for chemicals in ambient air. St. Paul, MN: Minnesota Department of Health, Environmental Health Division, www.health.state.mn.us/divs/eh/risk/rules/hrvrule.html.

NAS [1986]. Criteria and methods for preparing emergency exposure guidance level (EEGL), short-term public emergency guidance level (SPEGL), and continuous exposure guidance level (CEGL). National Academy of Science, Committee on Toxicology, Board of Environmental Studies and Toxicology, National Research Council. Washington, DC: National Academy Press, http://www.nap.edu/.

NAS [1996]. Toxicity of alternatives to chlorocarbons: HFC-134a and HCFC-123. National Academy of Science, Committee on Toxicology, Board of Environmental Studies and Toxicology, National Research Council. Washington, DC: National Academy Press, http://www.nap.edu/.

NAS [2001]. Standing operating procedures for developing acute exposure guideline levels for hazardous chemicals. National Academy of Science, Committee on Toxicology, Board of Environmental Studies and Toxicology, National Research Council. Washington, DC: National Academy Press, http://www.nap.edu/.

NAS [2002]. Final acute exposure guideline levels for selected airborne chemicals. Vol. 2. National Academy of Science, Committee on Toxicology, Board of Environmental Studies and Toxicology, National Research Council. Washington, DC: National Academy Press, www.nap.edu/.

NAS [2005]. Final acute exposure guideline levels (AEGLs) for acetone cyanohydrin, CAS# 75-86-5. National Academy of Science, Committee on Toxicology, Board of Environmental Studies and Toxicology, National Research Council. Washington, DC: National Academy Press, http://www.nap.edu/.

NAS [2007]. Interim acute exposure guideline levels (AEGLs) for selected metal phosphides. National Academy of Science, Committee on Toxicology, Board of Environmental Studies and Toxicology, National Research Council. Washington, DC: National Academy Press, http://www.nap.edu/.

NAS [2008]. Emergency and continuous exposure guidance levels for selected submarine contaminants. Vol. 2. National Academy of Science, Committee on Emergency and Continuous Exposure Guidance Levels for Selected Submarine Contaminants, Committee on Toxicology, National Research Council. Washington, DC: National Academy Press, http://www.nap.edu/.

NAS [2009]. Interim acute exposure guideline levels (AEGLs) for 25 selected chlorosilanes. National Academy of Science, Committee on Toxicology, Board on Environmental Studies and Toxicology, National Research Council. Washington, DC: National Academy Press, http://www.nap.edu/.

NASA [1999]. Spacecraft maximum allowable concentrations for airborne contaminants. JSC 20584. Houston, TX: National Aeronautics and Space Administration, Johnson Space Center, http://hefd.jsc.nasa.gov/toxeg.htm.

Nielsen GD, Wolkoff P, Alarie Y [2007]. Sensory irritation: risk assessment approaches. Regul Pharmacol Toxicol 48(1):6–18.

NIOSH [1987]. NIOSH respirator decision logic. Cincinnati, OH: U.S. Department of Health and Human Services, Centers for Disease Control, National Institute for Occupational Safety and Health. NIOSH Publication No. 87–108 (NTIS Publication No. PB-88-149612).

NIOSH [1994]. Documentation for immediately dangerous to life or health concentrations (IDLH). Cincinnati, OH: U.S. Department of Health and Human Services, Centers for Disease Control and Prevention, National Institute for Occupational Safety and Health. NTIS Publication No. PB-94-195047.

NIOSH [2004]. NIOSH respirator selection logic. Cincinnati, OH: U.S. Department of Health and Human Services, Centers for Disease Control and Prevention, National Institute for Occupational Safety and Health, NIOSH Publication No. 2005–100.

NIOSH [2005]. NIOSH pocket guide to chemical hazards. Cincinnati, OH: U.S. Department of Health and Human Services, Centers for Disease Control and Prevention, National Institute for Occupational Safety and Health. NIOSH Publication No. 2005–149.

NIOSH/OSHA [1981]. Occupational health guidelines for chemical hazards. Cincinnati, OH: U.S. Department of Health and Human Services, Centers for Disease Control and Prevention, National Institute for Occupational Safety and Health. NIOSH Publication No. 81–123 (NTIS Publication No. PB-83-154609).

NJ RTK [2010]. Right to know hazardous substance fact sheets. Trenton, NJ: Department of Health and Senior Services, New Jersey Right to Know, http://web.doh.state.nj.us/rtkhsfs/indexfs.aspx.

OSHA [2003]. Department of Labor: 29 CFR 1910.134. In: Code of Federal Regulations. Washington, DC: U.S. Government Printing Office, Office of the Federal Register.

Reinhardt CF, Azar A, Maxfield ME, Smith PE, Mullin LS [1971]. Cardiac arrhythmias and aerosol "sniffing." Arch Environ Health 22:265–279.

Rusch GM, Bst CB, Cavender FL [2009]. Establishing a point of departure for risk assessment using acute inhalation toxicity data. Regul Toxicol Pharmacol 54(3):247–255.

Schaper M [1993]. Development of a database for sensory irritants and its use in establishing occupational exposure limits. Am Ind Hyg Assoc J 54(9):488–544.

Shusterman D, Murphy MA, Balmes J [2003]. Differences in nasal irritant sensitivity by age, gender, and allergic rhinitis status. Int Arch Occup Environ Health 76:577–583.

TCEQ [2010]. Effects screening levels. Austin, TX: Texas Commission on Environmental Quality, Toxicology Branch, http://www.tceq.state.tx.us/nav/data/effectsscreeninglevels.html.

ten Berge WF, Zwart A, Appelman LM [1986]. Concentration-time mortality response relationship of irritant and systematically acting vapors and gases. J Haz Mat 13:301–309.

US DHS [2007]. Chemical facility anti-terrorism standards: interim final rule. U.S. Department of Homeland Security. Federal Register 72(67):17688–17745.

US DHS [2009]. Provisional advisory levels (PALs) for hazardous agents: research highlights. Washington, DC: U.S. Department of Homeland Security, Office of Research and Development, http://www.epa.gov/nhsrc/news/news121208.html.

US DOE [2008]. Temporary emergency exposure limits for chemicals: methods and practice. Washington, DC: U.S. Department of Energy, Technical Standards Program [DOE-HDBK-1046-2008], http://orise.orau.gov/emi/scapa/files/doe-hdbk-1046-2008_ac.pdf.

US DOT [2008]. Emergency response guidebook (ERG). Washington, DC: U.S. Department of Transportation, Homeland Security & Emergency Management Agency.

US EPA [1994] Methods for derivation of inhalation reference concentrations and application of inhalation dosimetry. EPA/600/8-90/066F. Washington, DC: U.S. Environmental Protection Agency.

US EPA [2010]. Integrated risk information system (IRIS): glossary. Washington, DC: U.S. Environmental Protection Agency, National Center for Environmental Assessment, http://www.epa.gov/iris/index.html.

van Raaij MTM, Janssen PAH, Piersma AH [2003]. The relevance of developmental toxicology endpoints for acute limit setting. The Netherlands: National Institute for Public Health and the Environment (RIVM). RIVM report 60190004/2003, http://www.epa.gov/opptintr/aegl/pubs/meetings/mtg35b.pdf.

Yant WP [1944]. Protecting workers against temporary and emergency exposures. In: Protecting plant manpower through the control of air contaminants. Special Bulletin No. 14. Washington, DC: U.S. Department of Labor, Division of Labor Standards.

Young RA, Bast CB, Wood CS, Adeshina F [2009]. Overview of the standing operating procedure (SOP) for the development of provisional advisory levels (PALs). Inhal Toxicol 21(Suppl)(3):1–11.

APPENDIX A: Example of the Derivation of an IDLH Value

This appendix illustrates the IDLH value derivation on the basis of the scientific criteria outlined in this document. This profile contains the proposed IDLH value and draft support documentation for chlorine (CAS# 7782-50-5). It should be noted that the information presented in this appendix is intended to serve only as an illustration of the application of the derivation process outlined in this CIB. For this reason, the proposed IDLH value for chlorine presented in this CIB should not be construed as official NIOSH policy.

A.1 Support Documentation for the Revised IDLH Value for Chlorine

Revised IDLH value: 2.8 ppm (8.1 mg/m^3)

Basis for revised IDLH value: The IDLH value is based on the human LOAEL for severe effects, on the basis of respiratory irritation and cough in human volunteers exposed to 2 ppm for 1 hour [Anglen 1981]. The LOAEL was adjusted to a 30-minute duration equivalent value of 3 ppm. A UF of 1 was applied to account for extrapolation from a threshold for severe effects in humans, yielding an IDLH value of 2.8 ppm rounded to 3 ppm.

1994 IDLH value: 10 ppm

Basis for 1994 IDLH value: The 1994 IDLH value for chlorine is 10 ppm, based on acute inhalation toxicity data in humans [NIOSH 1994].

NIOSH REL: 0.5 ppm (1.45 mg/m^3), 15-minute ceiling [NIOSH 2005]

Current OSHA PEL: 1 ppm (3 mg/m^3), ceiling [OSHA 2010]

1989 OSHA PEL:[‡] 0.5 ppm (1.5 mg/m^3), TWA; 1 ppm (3 mg/m^3), STEL [OSHA 1989]

2010 ACGIH TLV: 0.5 ppm, TWA; 1 ppm, STEL [ACGIH 2010]

2010 AIHA ERPG: ERPG-1: 1 ppm; ERPG-2: 3 ppm; ERPG-3: 20 ppm [AIHA 2010]

2010 AIHA WEEL: Not available

Description of substance: A greenish yellow gas with a pungent, irritating odor [NAS 2004]

LEL: Not available

NAC AEGL: Summarized in Table A–1. National Advisory Committee [2004] Final AEGLs: Chlorine.

A.2 Animal Toxicity Data

The available acute lethality database consists of multiple studies in mice [Lipton and Rotariu 1941; Silver et al. 1942; Schlagbauer and Henschler 1967; Bitron and Aharonson 1978; Alarie 1980; Jiang et al. 1983; O'Neil 1991], rats [Weedon et al. 1940; Back et al. 1972; MacEwen and Vernot 1972; Vernot et al. 1977; Zwart and Woutersen 1988], rabbits [Barrow and Smith 1975], and dogs [Underhill 1920; Withers and Lees 1985a]. The lowest LC$_{50}$ value in animals was identified from a study that exposed mice to 127 ppm for 30 minutes [Schlagbauer and Henschler 1967]. Critical animal lethality data are

---

[‡]1989 PELS are no longer legally enforced by federal OSHA, but many of these PELs were adopted by state OSHA plans, thus the 1989 PELs may still be in force in various states.

Table A–1. Summary of the AEGL values for chlorine

Classification	10 minutes	30 minutes	1 hour	4 hours	8 hours
AEGL-1	0.5 ppm	0.5 ppm	0.5 ppm	0.5 ppm	0.5 ppm
	1.5 mg/m^3	1.5 mg/m^3	1.5 mg/m^3	1.5 mg/m^3	1.5 mg/m^3
AEGL-2	2.8 ppm	2.8 ppm	2.0 ppm	1.0 ppm	0.71 ppm
	8.1 mg/m^3	8.1 mg/m^3	5.8 mg/m^3	2.9 mg/m^3	2.0 mg/m^3
AEGL-3	50.0 ppm	28.0 ppm	20.0 ppm	10.0 ppm	7.1 ppm
	145.0 mg/m^3	81.0 mg/m^3	58.0 mg/m^3	29.0 mg/m^3	21.0 mg/m^3

Abbreviation: mg/m^3 = milligrams per cubic meter; ppm = parts per million

Table A–2. Acute toxicity data and 30-minute-equivalent lethal concentration values for chlorine

Species	Reference	LC$_{50}$ (ppm)	Time (minutes)	Adjusted 30-minute LC*	UF[†]	30-minute derived value (ppm)[‡]
Mouse	Schlagbauer and Henschler [1967]	127	30	127	30	4.2
Mouse	Zwart and Woutersen [1988]	504	30	504	30	17
Rat	Zwart and Woutersen [1988]	703	30	703	30	23

Abbreviation: LC = lethal concentration; LC$_{50}$ value = median lethal concentration; ppm = parts per million; UF = uncertainty factor.
*No time adjustment was made to LC$_{50}$ values.
[†]The selection of the UF for chlorine was based on Chapter 4.0: Use of Uncertainty Factors. The UF of 30 was selected on the basis of (1) the extrapolation from a concentration that is lethal to animals, (2) animal to human differences, and (3) human variability.
[‡]Derived values are calculated by dividing the Adjusted 30-minute LC by the UF.

Table A–3. Acute toxicity data and 30-minute-equivalent non-lethal concentration values for chlorine

Species	Reference	LOAEL (ppm)	Time (minutes)	Adjusted 30 minute LC*	UF[†]	30-minute derived value (ppm)[‡]
Mouse	Jiang et al. [1983]	9.1	360	32	10	3.2
Rat	Jiang et al. [1983]	9.1	360	32	10	3.2

Abbreviation: LOAEL = lowest observed adverse effect level; ppm = parts per million; UF = uncertainty factor.
*For exposures other than 30 minutes, the ten Berge et al. [1986] relationship is used for duration adjustment ($C^n \times t = k$); no empirically estimated n values were available; therefore, the default values were used: $n = 3$ for exposures greater than 30 minutes and $n = 1$ for exposures less than 30 minutes.
[†]The selection of the UF for chlorine was based on Chapter 4.0: Use of Uncertainty Factors. The UF of 10 was selected on the basis of (1) animal to human differences, and (2) human variability.
[‡]Derived values are calculated by dividing the Adjusted 30-minute LC by the UF.

summarized in Table A–2, along with time adjustments, UF, and potential derived IDLH values.

Non-lethal effects in mice, rabbits, and rats consisted of ocular and nasal irritation, transient changes in lung function, bronchitis, lesions in the nasal passages and lung, and mild edema [Barrow and Smith 1975; Jiang et al. 1983; Zwart and Woutersen 1988; Demnati et al. 1995]. Nasal lesions appear to be the most sensitive effect, as they occur at the lowest tested concentrations of chlorine in both rats and mice [Jiang et al. 1983]. Multiple RD_{50} studies have also been conducted in mice [Barrow et al. 1977; Barrow and Steinhagen 1982; Chang and Barrow 1984; Gagnaire et al. 1994]. Critical animal non-lethality data are summarized in Table A–3 along with time adjustments, UFs, and potential derived IDLH values.

A.3 Human Data

Deaths have been reported after inhalation exposures to chlorine, but specific exposure concentrations are not available from reports of accidental releases. Withers and Lees [1985b] estimated lethal concentrations to humans using a probit analysis of available information. They estimated a 30-minute LC_{50} value and LC_{10} value of 100 and 50 ppm, respectively, for vulnerable populations. Critical human lethality data are summarized in Table A–4, along with time adjustments, UFs, and potential derived IDLH values.

Experimental exposure to non-lethal concentrations of chlorine has caused changes in nasal air resistance [D'Alessandro et al. 1996; Shusterman et al. 1998], transient changes in pulmonary function [Anglen 1981; Rotman et al. 1983; D'Alessandro et al. 1996], irritation [Joosting and Verberk 1974; Anglen 1981; Rotman et al. 1983], and cough [Joosting and Verberk 1974; Anglen 1981]. The lowest concentration at which mild discomfort due to irritation or cough was reported is at 1 ppm for durations of 4 hours or greater [Anglen 1981; Rotman et al. 1983]; however, irritation was not significant among volunteers exposed to 2 ppm for 30 minutes [Anglen 1981].

Multiple accidental exposures to non-lethal concentrations of chlorine have been reported, but specific exposure conditions and durations are not available. Symptoms of dyspnea, cough, and irritation are most commonly reported in the literature [Shroff et al. 1988; Abhyankar et al. 1989; Mrvos et al. 1993; ILO 1998], but other severe systemic effects (such as headache, vomiting, giddiness, chest

Table A–4. Acute toxicity data and 30-minute-equivalent lethal concentration values for chlorine

Species	Reference	LC_{LO} (ppm)	Time (minutes)	Adjusted 30-minute LC*	UF[†]	30-Minute derived value (ppm)[‡]
Human	Withers and Lees [1985b]	50	30	50	3	17

Abbreviation: LC = Lethal concentration; LC_{LO} = lowest lethal concentration of a substance in the air reported to cause death; ppm = parts per million; UF = uncertainty factor.
*No time adjustment was made to the LC_{LO} value.
[†]The selection of the UF for chlorine was based on Chapter 4.0: Use of Uncertainty Factors. The UF of 3 was selected on the basis of human variability.
[‡]Derived values are calculated by dividing the Adjusted 30-minute LC by the UF.

Table A–5. Acute toxicity data and 30-minute-equivalent non-lethal concentration values for chlorine

Species	Reference	NOAEL (ppm)	LOAEL (ppm)	Time (minutes)	Adjusted 30-minute LC*	UF†	30-minute derived value (ppm)‡
Human	Anglen [1981]	2.0		30	2.0	1	2.0
Human	Anglen [1981]		2.0	60	2.8	1	2.8
Human	Rotman et al. [1983]		1.0	240	2.8	1	2.8
Human	ILO [1971]		40.0	30	40	3	13

Abbreviation: LC =lethal concentration; LOAEL = lowest observed adverse effect level; NOAEL = no observed adverse effect level; ppm = parts per million; UF = uncertainty factor

*For exposures other than 30 minutes, the ten Berge et al. [1986] relationship is used for duration adjustment ($C^n \times t = k$); no empirically estimated n values were available; therefore, the default values were used: $n = 3$ for exposures greater than 30 minutes and $n = 1$ for exposures less than 30 minutes.

†The selection of the UF for chlorine was based on *Chapter 4: Use of Uncertainty Factors*. The UF of 1 or 3 was selected on the basis of (1) severe effects or (2) human variability.

‡Derived values are calculated by dividing the Adjusted 30-minute LC by the UF.

pain, and abdominal discomfort) have also been reported [Shroff et al. 1988; Abhyankar et al. 1989]. Critical human non-lethality data are summarized in Table A–5, along with time adjustments, UFs, and potential derived IDLH values.

A.4 IDLH Value Rationale Summary

Among the acute lethality studies, the mouse provides the lowest LC_{50} value of 127 ppm for a 30-minute exposure period [Schlagbauer and Henschler 1967]. A UF of 30 was applied to account for extrapolation from a concentration that is lethal to animals, animal to human differences, and human variability, resulting in a potential IDLH value of 4.2 ppm. Anglen [1981] reported a LOAEL of 2 ppm for throat irritation and cough in human volunteers exposed to chlorine for 1 hour. The LOAEL was duration-adjusted to a 30-minute-equivalent value of 2.8 ppm. A safety factor of 1 was applied to account for extrapolation from a threshold for irritant effects in humans, resulting in an IDLH value of 3 ppm rounded to 3 ppm.. This value is supported by the presence of irritant effects in human volunteers exposed to 1 ppm of chlorine for 4 hours [Anglen 1981; Rotman et al. 1983], which would also result in a similar IDLH value.

Appendix A: References

Abhyankar AN, Bhambure N, Kamath NN, Pajankar SP, Nabar ST, Shrenivas A, Shah AC, Deshmukh SN [1989]. Six month follow-up of fourteen victims with short-term exposure to chlorine gas. J Soc Occup Med 39:131–132.

ACGIH [2010]. TLVs® (threshold limit values) and BEIs® (biological exposure indices). Cincinnati, OH: American Conference of Governmental Industrial Hygienists Signature Publications.

AIHA [2010]. Annual ERPG (emergency response planning guidelines) and WEEL (workplace environmental exposure levels) handbook. Fairfax, VA: American Industrial Hygiene Association Press.

Alarie Y [1980]. Toxicological evaluation of airborne chemical irritants and allergens using respiratory reflex reactions. In: Proceedings, Symposium on Inhalation Toxicology and Technology, Ann Arbor, MI: Ann Arbor Science, pp 207–231.

Anglen DM [1981]. Sensory response of human subjects to chlorine in air [dissertation]. Ann Arbor, MI: University of Michigan.

Back KC, Thomas AA, MacEwen JD [1972]. Reclassification of materials listed as transportation health hazards. Report no. TSA-20-72-3. Dayton, OH: Aerospace Medical Research Laboratory, Wright-Patterson AFB.

Barrow CS, Smith RG [1975]. Chlorine induced pulmonary function changes in rabbits. Am Ind Hyg Assoc J 36:398–403.

Barrow CS, Steinhagen WH [1982]. Sensory irritation tolerance development to chlorine in F-344 rats following repeated inhalation. Toxicol Appl Pharmacol 65:383–389.

Barrow CS, Alarie Y, Warrick JC, Stock MF [1977]. Comparison of the sensory irritation response in mice to chlorine and hydrogen chloride. Arch Environ Health 32:68–76.

Bitron MD, Aharonson EF [1978]. Delayed mortality of mice following inhalation of acute doses of formaldehyde, sulfur dioxide, chlorine and bromine. Am Ind Hyg Assoc J 39:129–138.

Chang JCF, Barrow CS [1984]. Sensory tolerance and cross-tolerance in F-344 rats exposed to chlorine or formaldehyde gas. Toxicol Appl Pharmacol 76:319–327.

D'Alessandro A, Kuschner W, Wong H, Boushey HA, Blanc PD [1996]. Exaggerated responses to chlorine inhalation among persons with nonspecific airway hyperreactivity. Chest 109:331–337.

Demanti R, Fraser R, Plaa G, Malo JL [1995]. Histopathological effects of acute exposure to chlorine gas on Sprague-Dawley rat lungs. J Environ Pathol Toxicol Oncol 14:15–19.

Gagnaire F, Azim S, Bonnet P, Hecht G, Hery M [1994]. Comparison of the sensory irritation response in mice to chlorine and nitrogen trichloride. J Appl Toxicol 14:405–409.

ILO [1971]. Chlorine and compounds. In: Encyclopaedia of occupational health and safety. 4th ed. Vol. 1 (A–K). Geneva, Switzerland: International Labour Office.

ILO [1998]. Chlorine and compounds. In: Encyclopaedia of occupational health and safety. 4th ed. Vol. 1 (A–K). Geneva, Switzerland: International Labour Office.

Jiang XZ, Buckley LA, Morgan KT [1983]. Pathology of toxic responses to the RD_{50} concentration of chlorine gas in the nasal passages of rats and mice. Toxicol Appl Pharmacol 71:225–236.

Joosting P, Verberk M [1974]. Emergency population exposure: a methodological approach. Recent Adv Assess Health Eff Environ Pollut 4:2005–2029.

Lipton MA, Rotariu GJ [1941]. In: EMK Gelling, McLean FC, eds. Progress report on toxicity of chlorine gas for mice. Report no. 286. Washington, DC: U.S. National Defense Committee, Office of Science Research and Development.

MacEwen JD, Vernot EH [1972]. Toxic hazards research unit annual technical report: 1972. AMRL-TR-72-62. Dayton, OH: Aerospace Medical Research Laboratory, Wright-Patterson Air Force Base; Springfield, VA: National Technical Information Service.

Mrvos R, Dean BS, Krenzelok EP [1993]. Home exposures to chlorine/chloramine gas: review of 216 cases. South Med J 86:654–657.

NAS [2004]. Interim acute exposure guideline levels (AEGLs): chlorine, CAS# 7782-50-5. Washington, DC: National Academy of Sciences, Commission on Life Sciences, National Research Council.

NIOSH [1994]. Documentation for Immediately Dangerous to Life or Health Concentrations (IDLHs): Chlorine. Cincinnati, OH: Centers for

Disease Control and Prevention, National Institute for Occupational Safety and Health, http://www.cdc.gov/niosh/idlh/7782505.html.

NIOSH [2005]. NIOSH pocket guide to chemical hazards. NIOSH publication no. 2005-149. Cincinnati, OH: Centers for Disease Control and Prevention, National Institute for Occupational Safety and Health, http://www.cdc.gov/niosh/npg/default.html.

O'Neil CE [1991] Immune responsiveness in chlorine exposed rats. PB92-124478. Cincinnati, OH: Centers for Disease Control and Prevention, National Institute for Occupational Safety and Health.

OSHA [1989]. OSHA PEL project documentation, 1988. NIOSH databases, http://www.cdc.gov/niosh/pel88/pelstart.html.

OSHA [2010]. Occupational Safety and Health Standards. 29 CFR 1910. Subpart Z: Toxic and Hazardous Substances. Washington, DC: Occupational Safety and Health Administration, http://www.osha.gov/pls/oshaweb/owadisp.show_document?p_table=standards&p_id=9992.

Rotman HH, Fliegelman MJ, Moore T, Smith RG, Anglen DM, Kowalski CJ, Weg JG [1983]. Effects of low chlorine and bromine concentrations on man. Int Arch Gewerbepathol 23:79–90.

Schlagbauer M, Henschler D [1967]. Toxicity of chlorine and bromine with single and repeated inhalation. Int Arch Gewerbepathol Gewerbehyg 23:91.

Shroff CP, Khade MV, Srinivasan M [1988]. Respiratory cytopathology in chlorine gas toxicity: a study in 28 subjects. Diagn Cytopathol 4:28–32.

Shusterman DJ, Murphy MA, Balmes JR [1998]. Subjects with seasonal allergic rhinitis and nonrhinitic subjects react differentially to nasal provocation with chlorine gas. J Allergy Clin Immunol 101:732–740.

Silver SD, McGrath FP, Ferguson RL [1942]. Chlorine median lethal concentration data for mice. DATR 373. Aberdeen, MD: Edgewood Arsenal.

ten Berge WF, Zwart A, Appelman LM [1986]. Concentration-time mortality response relationship of irritant and systematically acting vapors and gases. J Haz Mat 13:301–309.

Underhill FP [1920]. The lethal war gases: physiology and experimental treatment. New Haven, CT: Yale University Press, p. 20.

Vernot EH, MacEwen JD, Haun CC, Kinkead ER [1977]. Acute toxicity and skin corrosion data for some organic and inorganic compounds and aqueous solutions. Toxicol Appl Pharmacol 42:417–423.

Weedon FR, Hartzell A, Setterstrom C [1940]. Toxicity of ammonia, chlorine, hydrogen cyanide, hydrogen sulfide and sulfur dioxide gases. V: Animals. Contrib Boyce Thompson Inst 11:365–386.

Withers RMJ, Lees FP [1985a]. The assessment of major hazards: the lethal toxicity of chlorine. Part 1: review of information on toxicity. J Hazard Mater 12:231–282.

Withers RMJ, Lees FP [1985b]. The assessment of major hazards: the lethal toxicity of chlorine. Part 2: model of toxicity to man. J Hazard Mater 12:283–302.

Zwart A, Wouterson RA [1988]. Acute inhalation toxicity of chlorine in rats and mice: time-concentration-mortality relationships and effects on respiration. J Hazard Mater 19:195–208.

APPENDIX B: IDLH Value Development Prioritization

This appendix identifies how NIOSH will determine the priorities for developing IDLH values. The guidance values play an important role in planning work practices surrounding potential high-exposure environments in the workplace and in guiding actions by emergency response personnel during unplanned exposure events. Ideally, IDLH values would be available for all chemicals that might be present under high exposure situations. However, this breadth of coverage of IDLH values is not practical and might not even be necessary for many chemicals, such as those with very low exposure potential or those that are not acutely toxic. In addition, the absence of data and limited resources makes it difficult to evaluate the multitude of chemicals currently available in commerce. Therefore, a prioritization process is used by NIOSH to ensure that resources are allocated to yield the greatest impact on risk reduction in the event that control measures fail (including respiratory protection devices). This process takes into account both toxicity and exposure potential, and it is applied to a broad pool of relevant chemicals (e.g., chemical warfare agents, industrial chemicals, high-production-volume [HPV] chemicals, or agrochemicals subject to emergency or uncontrolled releases). A qualitative algorithm is used to generate a tentative relative priority ranking. This process is intended only to provide tentative guidance based on a simple approach that uses readily available sources of information. The resulting priorities are further modified on the basis of NIOSH emphasis areas. For example, chemicals can be added or removed from the list on the basis of new information related to toxicity or exposure potential. The development and use of a documented prioritization process allows for frequent updating of both input data and prioritization criteria to meet changing needs.

Substances considered in the ranking process are compiled from existing databases of chemicals identified by other agencies as "of concern" because of use in chemical terrorism or as chemicals with the potential for exposure due to other uncontrolled releases (and thus having greater opportunities for high, acute exposures). Existing lists of agents of concern may not be fully representative of industrial chemicals for which acute exposures may occur during planned activities (e.g., special maintenance activities) or unplanned-release events. However, IDLH values for many of these sorts of chemicals were included in the original IDLH value development process and in the 1994 updates. Moreover, NIOSH adds additional chemicals of interest that are nominated by interested stakeholders or the subject of new emphasis programs. Chemicals from the following databases (as supplemented by NIOSH chemicals of interest) were included in the ranking process:

- **Hazardous Substances Emergency Events Surveillance (HSEES)**—This database contains self-reported incidents of accidental chemical releases. The database was created by the Agency for Toxic Substances and Disease Registry (ATSDR) [ATSDR 2008].

- **Emergency Preparedness and Response**—This list of specific agents and other threat agents was created by the Centers for Disease Control and Prevention (CDC) [CDC 2008].

- **Emergency Response Guidebook (ERG)**—This list of toxic by inhalation (TIH) chemicals and water-reactive TIH chemicals was created by the DOT [US DOT 2008].

- **Chemical Facility Anti-Terrorism Standards (CFATS), Appendix A**—This list of chemicals of interest to national security was created by the Department of Homeland Security (DHS) [US DHS 2007].

Exposure-related parameters can be divided into two categories: 1) those that provide a direct indication of exposure potential (e.g., number of recorded accidents or spills involving a chemical) and 2) data that provide indirect indication of exposure potential (e.g., volume produced). In weighing such metrics, a balance needs to be struck between the greater confidence provided by direct-release data based on the obvious relevance to exposure potential, and the need to have data on exposure potential that are available for most chemicals. Information on direct exposure indicators was obtained from the HSEES database [ATSDR 2008]. Although only 14 states participate in the program, the data are useful as an exposure indicator. Evidence of frequent past incidents involving uncontrolled releases receives a score of 1, and the absence of reporting of prior releases is scored 0.

Chemical production volume is used as an indirect indication of exposure potential [USEPA 2008]. The USEPA classifies HPV chemicals as those chemicals produced or imported in the United States in quantities of 1 million pounds or more per year; medium-production chemicals are quantities of 25,000 to less than 1 million pounds per year; and low-production chemicals are quantities less than 25,000 pounds per year. HPV chemicals receive a score of 1, whereas low-and medium-production-volume chemicals receive a score of 0.

Because the aim of the prioritization process is the development of guidance for protection from acute inhalation exposures, endpoints that best inform the potential for life-threatening, irreversible, or escape-impairing effects following acute inhalation exposures receive the greatest weight. The following approach and resources are used to score toxicity considerations:

1. Direct indication of exposure potential (e.g., number of recorded accidents or spills involving a chemical).

   - Evidence of frequent past incidents involving uncontrolled releases
   - HSEES—collects and analyzes actual hazardous chemical releases and emergency responder injuries
   - Chemicals with uncontrolled releases (URs) are scored as a 1, and lack of reported data is scored as a 0.

2. Indirect indication of exposure potential (e.g., volume produced)

   - Indicative of the potential for exposure from the amount of chemical that is produced
   - USEPA classifies chemicals as low, medium, or high production volume (HPV)
   - Chemicals classified as HPV are scored as a 1, whereas low- and medium-volume chemicals are scored as a 0.

3. Short-term exposure limits (STELs)—NIOSH RELs, OSHA PELs, AIHA WEELs, and ACGIH TLVs® [ACGIH 2008; AIHA 2008; NIOSH 2007]

   - STEL values below 20 ppm for vapors and gases or 2 mg/m^3 for particulates provide a reasonable cut point for identifying the most significantly acutely toxic substances.
   - Substances with a STEL below these cut points receive a score of 1, whereas substances with a STEL equal to or greater than these values or that have no available STEL receive a score of 0.

4. Irritant Potential (IRR)—*NIOSH Pocket Guide to Chemical Hazards* [NIOSH 2005] or the European Union (EU) Risk Phrases (R-phrases) [EU 2008] for irritation

   - Irritants receive a score of 0.5 and corrosive chemicals receive a score of 1; all other chemicals receive a score of 0.

5. Acute toxicity (AT) (e.g., lethal concentration resulting in 50% mortality in exposed animals [LC50])—*Registry of Toxic Effects of Chemical Substances* (RTECS) [CCOHS 2008]

   - Chemicals classified as extremely or highly hazardous in RTECS or with an EU R-phrase of "very toxic" or "toxic" are scored as 1; otherwise, chemicals are scored as 0.

- Chemicals that have not been evaluated by means of these systems are judged on the basis of the lowest reliable LC_{50} compared to the EU R-phrase criteria.

6. Developmental toxicant (DT)—*NIOSH Pocket Guide to Chemical Hazards* [NIOSH 2005] or California Environmental Protection Agency (Cal/EPA) Proposition 65 list [Cal/EPA 2008].

    - Chemicals identified as reproductive/developmental toxicants are scored as 0.5; otherwise, chemicals are scored 0.

7. Carcinogenicity (CA)—EPA, International Agency for Research on Cancer (IARC) [IARC 2008], ACGIH [2008], *NIOSH Pocket Guide to Chemical Hazards* [NIOSH 2005], Cal/EPA Proposition 65 List [Cal/EPA 2008], or other sources

    - Chemicals classified by recognized systems as probable, likely, or known human carcinogens are scored as 0.5; otherwise, chemicals are scored as 0.

8. Other considerations are used qualitatively to further refine priorities among chemicals with the same risk-based score. These Tier II considerations include the following:

    - Availability of other acute exposure guidance—Such guidance includes existing IDLH, AEGL, or ERPG values. The availability of such guidance decreases the urgency for developing (or revising) IDLH values.

    - Availability of toxicity data—the absence of adequate data precludes the development of an IDLH value. The lack of toxicity data for a chemical with high exposure potential is used to identify research needs.

    - Availability of exposure monitoring methods—The availability of a validated sampling and analytical method increases the likely near-term utility of a derived IDLH value. The absence of a validated sampling and analytical method for high-priority chemicals could be used to identify research needs.

    - Presence on existing lists of high priority agents—If other agencies have listed the material as a high priority, then the IDLH value may be useful to other agencies. This type of leveraging of resources is desirable and also helps to harmonize levels of worker health protection among agencies with related missions.

    - Degree of safety hazard—If potential risk for two or more chemicals as determined on the basis of chemical toxicity is equal, then agents that have a greater degree of safety-related risk (e.g., flammability) are given greater weight. This consideration allows for easier comparison of overall risk profiles and selection of the most appropriate basis for risk management (e.g., developing entry criteria or emergency plans on the basis of whichever is the greater concern, safety or health risk).

The overall priority score is the sum of the exposure score and toxicity score:

- Tier I: Risk Priority Score = Exposure Score [ranges from 0 to 2] + Toxicity Score [ranges from 0 to 3]

    Risk Priority Score =
    [UR + PV] + [STEL + IRR + AT + DT + CA]

    Where:
    AT = acute toxicant
    CA = carcinogenicity
    DT = developmental toxicant
    IRR = irritant
    PV = production volume
    STEL = short-term exposure limit
    UR = uncontrolled release

- Tier II: Used qualitatively to make an overall judgment on priorities among chemicals with the same risk priority score.

Appendix B: References

ACGIH [2008]. Threshold limit values (TLVs®) and biological exposure indices (BEIs®). Cincinnati, OH: American Conference of Governmental Industrial Hygienists.

AIHA [2008]. The emergency response planning guidelines (ERPG) and workplace environmental exposure levels (WEEL) handbook. Fairfax, VA: American Industrial Hygiene Association.

ATSDR [2008]. Hazardous substances emergency events surveillance (HSEES) system. Atlanta, GA: Department of Health and Human Services, Centers for Disease Control and Prevention, Agency for Toxic Substances and Disease Registry, http://www.atsdr.cdc.gov/HS/HSEES/index.html.

Cal/EPA [2008]. Proposition 65: chemicals known to the state to cause cancer or reproductive toxicity. August 1, 2008. Sacramento, CA: California Environmental Protection Agency, Office of Environmental Health Hazard Assessment, http://www.oehha.org/prop65/prop65_list/files/P65single080108.pdf.

CCOHS [2008]. RTECS: Registry of Toxic Effects of Chemical Substances. Ontario, Canada: Canadian Centre for Occupational Health and Safety.

CDC [2008]. Emergency preparedness and response chemical emergencies: agents, diseases and other threats. Atlanta, GA: Department of Health and Human Services, Centers for Disease Control and Prevention, http://www.bt.cdc.gov/chemical/.

EU [2008]. ESIS (European Chemical Substances Information System): EU risk phrases. Brussels, Belgium: European Union, Joint Research Centre, European Commission, http://ecb.jrc.it/esis/.

IARC [2008]. IARC monographs on the evaluation of carcinogenic risks to humans. Lyon, France: International Agency for Research on Cancer, http://monographs.iarc.fr/.

NIOSH [2005]. Pocket guide to chemical hazards. Cincinnati, OH: U.S. Department of Health and Human Services, Centers for Disease Control and Prevention, National Institute for Occupational Safety and Health, http://www.cdc.gov/niosh/npg/default.html.

US DHS [2007]. Chemical facility anti-terrorism standards: appendix A. Washington, DC: U.S. Department of Homeland Security, http://www.dhs.gov/xlibrary/assets/chemsec_appendixa-chemicalofinterestlist.pdf.

US DOT [2008]. Emergency response guidebook. Washington, DC: U.S. Department of Transportation, Office of Hazardous Materials Initiatives and Training, Pipeline and Hazardous Materials Safety Administration, http://hazmat.dot.gov/pubs/erg/erg2008_eng.pdf.

US EPA [2008]. High production volume list. Prevention, pesticides and toxic substances, pollution prevention and toxics. Washington, DC: U.S. Environmental Protection Agency, http://www.epa.gov/chemrtk/pubs/general/opptsrch.htm.

APPENDIX C: Critical Effect Determination for IDLH Value Development—Consideration of Severity, Reversibility, and Impact on Escape Impairment

As discussed in the main document, the intent of the IDLH value is to protect against exposures that are "likely to cause death or immediate or delayed permanent adverse health effects or prevent escape from such an environment." In other words, the most appropriate effects to use as the basis of the IDLH value derivation are those that are severe, irreversible, or escape-impairing. Scientific judgment is an important aspect in evaluating severity of effects and determining which ones are irreversible, but guidance is available from a number of different sources.

Severe adverse effects that are not necessarily immediately escape-impairing are judged on a case-by-case basis weighing considerations such as the need for medical treatment, the potential for altered function or disability, and the potential for long-term deficits in function. These include severe, but reversible, acute effects such as hemolysis, chemical asphyxia, delayed pulmonary edema, and significant acute organ damage (hepatitis, decreased kidney function, etc.). If these effects could be caused by the chemical, it is important that the available toxicity studies evaluated the development of such effects by, for example, allowing sufficient time between exposure and evaluation of the endpoint.

Guidance on evaluating and ranking the severity of toxic effects is available from a number of organizations. DeRosa et al. [1985] developed a 10-category scheme for evaluating noncancer toxicity in the evaluation of Reportable Quantities under the USEPA Superfund legislation. Although designed for the context of chronic exposures, this approach provides insight into the relative severity of different types of histopathology and developmental toxicity. The Agency for Toxic Substances and Disease Registry (ATSDR) includes the following five severity rankings [Pohl and Abadin 1995]:

- No Observed Effect Level (NOEL)
- No Observed Adverse Effect Level (NOAEL)
- Minimal Lowest Observed Adverse Effect Level ($LOAEL_1$)
- Moderate Lowest Observed Adverse Effect Level ($LOAEL_2$)
- Frank Effect Level (FEL)

ATSDR applies this approach from acute exposures (defined as exposures up to 14 days) through chronic exposures, and a number of publications are available on applying this approach to various types of effects (e.g., Abadin et al. 1998, 2007; Chou and Pohl 2005; Pohl and Chou 2005; Pohl et al. 2005). Although intended for a different purpose, these analyses can provide insights into the evaluation of effect severity. In particular, the "moderate" LOAEL category used by ATSDR is more likely to be considered severe or irreversible, and thus relevant to IDLH value development. Guidance on evaluation of the severity of effects is also available from the USEPA RfC guidelines [USEPA 1994] and from the American Thoracic Society (e.g., Pellegrino et al. 2005).

Determining which effects are escape-impairing is complicated both by the limited guidance available from other sources and by the fact that reporting of signs and symptoms for similar underlying effects may differ across human and animal studies. For example, the same underlying mechanism may be described as inducing intolerable irritation in a human clinical study or case report, but may manifest as changes in respiration rate, nasal discharge, or altered activity level in an acute toxicity test in animals. For this reason, guidance was developed that allows for more consistent assigning of comparative severity of observed effects (i.e., escape-impairing versus non-escape-impairing) for commonly observed adverse effects used as the basis of IDLH values. Table C–1 provides guidance for classification of many effects commonly seen in acute studies. Because of the nature of the evaluation methods, endpoints that can be evaluated in humans are generally limited to clinical signs and symptoms, along with some specialized testing, and some histopathology evaluation that can be conducted non-invasively (e.g., for the nasal cavity), or can be inferred from other evaluations (e.g., pulmonary edema).

Table C–1. Common clinical signs, symptoms, and histopathological abnormalities observed during acute exposures

Clinical sign(s)	Escape-impairing?	Humans	Animals	Comments
Irritation—Ocular				
Signs and symptoms—Ocular				
Eye irritation (subjective description)	Yes	X		If moderate or severe
Lacrimation (excessive tearing, clear or colored)	Yes	X	X	If severe (assumes will be accompanied by other severe irritation responses)
Blepharospasm (eye squinting and shutting)	Yes	X	X	If severe
Reduced or poor vision	Yes	X		If severe
Mouth- or face-pawing activity	No		X	May be observed even during mild irritation
Eye blink rate/frequency	No	X	X	Difficult to use as a correlate of irritation, although some investigators assert that it may be useful as a marker of moderate to severe eye irritation
Ocular examination findings				
Swelling of eyelids	Yes	X	X	If eyelids are closed (more than half-closed)
Scattered or diffuse areas of opacity (other than slight dulling of normal luster), details of iris clearly visible	No	X	X	Assuming no significant impairment of vision
Easily discernible translucent area, details of iris slightly obscured	No	X	X	
Nacreous (lustrous) area, no details or iris visible, size of pupil barely discernible	Yes	X	X	
Opaque cornea, iris not discernible through the opacity	Yes	X	X	If severe, assumes will significantly impair vision

(Continued)

Table C–1 (Continued). Common clinical signs, symptoms, and histopathological abnormalities observed during acute exposures

Clinical sign(s)	Escape-impairing?	Humans	Animals	Comments
Markedly deepened rugae (folds or wrinkles), congestion, swelling, moderate circumcorneal hyperemia, or injection, or any of these or combination of any thereof, iris still reacting to light (sluggish reaction is positive)	Yes	X	X	
No reaction to light, hemorrhage, gross destruction (any or all of these)	Yes	X	X	
Ocular hyperemia (blood vessels hyperemic causing red eye)	No	X	X	
Cornea, inflammation or abrasion	Yes	X	X	
Cataract	Yes	X	X	
Inflammation of the eyes	Yes	X	X	If inflammation severe, assumes this correlates to severe irritation; large changes in some sensitive biomarkers may not necessarily indicate severe irritation responses.

Irritation - Respiratory

Nasal signs and symptoms

Clinical sign(s)	Escape-impairing?	Humans	Animals	Comments
Nasal irritation or pain (subjective description)	No	X		
Nasal localization	No	X		Endpoint differentiates sharp smell from irritation
Sneezing	Yes	X	X	If severe
Nasal congestion	No	X	X	
Nostril discharges: red or colorless	No	X	X	

(Continued)

Table C–1 (Continued). Common clinical signs, symptoms, and histopathological abnormalities observed during acute exposures

Clinical sign(s)	Escape-impairing?	Humans	Animals	Comments
Thickness/swelling of nasal mucosa (decreased nasal cross-sectional area)	No	X	X	Methods measuring mucosal thickness not directly related to sensory irritation effects
Increased nasal airway resistance	No	X		

Respiratory tract symptoms

Clinical sign(s)	Escape-impairing?	Humans	Animals	Comments
Dry cough	Yes	X		If severe
Cough with mucus or blood	Yes	X		If severe
Chest wheezing	Yes	X		If severe, assumes may impair breathing
Rales (rapid series of short, loud sounds)	No	X		
Breathing rate/volume measured by pulmonary function test results	Yes	X	X	If severe in humans, assumes may impair breathing; concentrations in the range of the RD_{50} in rodents
Dyspnea (difficult or labored breathing observed as abdominal breathing or gasping)	Yes	X		If severe in humans, assumes may impair breathing
Painful breathing	Yes	X		If severe in humans, assumes sufficient to impair breathing
Apnea (a transient cessation of breathing following a forced respiration)	Yes	X	X	Indication of sufficient irritation to modify breathing
Tachypnea (quick and usually shallow respiration)	Yes	X	X	Indication of sufficient irritation to modify breathing
Cyanosis (bluish appearance of tail, mouth, foot pads, skin or mucous membranes)	Yes	X	X	If severe, assumes sufficient to impair respiration
Laryngoconstriction	Yes	X	X	If severe, assumes may impair breathing
Bronchoconstriction	Yes	X	X	If moderate or severe, assumes may impair breathing

(Continued)

Table C–1 (Continued). Common clinical signs, symptoms, and histopathological abnormalities observed during acute exposures

Clinical sign(s)	Escape-impairing?	Humans	Animals	Comments
Respiratory tract histopathology				
Nasopharynx inflammation	No	X	X	
Nasopharynx erosion or necrosis	No	X	X	
Larynx inflammation	Yes	X	X	If severe, assumes may impair breathing
Larynx erosion or necrosis	Yes		X	If moderate or severe, assumes may impair breathing
Tracheal or bronchial inflammation	Yes		X	If severe, assumes may impair breathing
Tracheal or bronchial erosion or necrosis	Yes		X	If moderate or severe, assumes may impair breathing
Alveolar hemorrhage or necrosis	Yes		X	If observed, assumes may impair breathing
Pulmonary edema	Yes	X	X	If observed, assumes may impair breathing
Neurological				
Signs and symptoms				
Arousal state	No	X	X	If sluggish with some exploratory movements with periods of immobility, or hyperalert, excited, sudden bouts of running or body movements, or changes in rearing
Headache	Yes	X		Only if described in study as debilitating
Lightheadedness, dizziness/faintness	Yes	X		If moderate or severe
Lassitude/lethargy (feeling low in energy or slowed)	No	X		Assume if severe, lassitude would be seen as extreme drowsiness/fatigue
Extreme drowsiness, fatigue or sleepiness (somnolence)	Yes	X		If severe
Narcosis (stupor)	Yes	X	X	If observed

(Continued)

Table C–1 (Continued). Common clinical signs, symptoms, and histopathological abnormalities observed during acute exposures

Clinical sign(s)	Escape-impairing?	Humans	Animals	Comments
Frank effects (including postural observations—excessive sway, lying on side, limbs in the air, loss of balance, stupor, convulsions, seizure, coma)	Yes	X	X	If observed
Exhilaration (unusual)	No	X		
Euphoria (a feeling of exaggerated elation)	No	X		
Loss of concentration	Yes	X	X	If severe
Loss of recent memory	Yes	X	X	
Long-term memory loss	No	X	X	
Unstable moods	No	X		
CNS excitability				
Clonic movements (marked by alternate contraction and relaxation of muscles)	Yes	X	X	If moderate-severe body tremors and myoclonic jerks
Tonic movements (marked by continuous muscular contractions)	Yes	X	X	If head and body rigidly forward or backward
Autonomic effects				
Palpebral closure, ptosis or relaxation of nictating membranes	Yes	X	X	If eyelids or nictating membranes drooping; drooping of nictating membranes would not be observed in humans
Urination	Yes	X	X	Common effects of nerve agents and accompanied by changes that impair escape
Defecation	Yes	X	X	Common effects of nerve agents and accompanied by changes that impair escape

(Continued)

Table C–1 (Continued). Common clinical signs, symptoms, and histopathological abnormalities observed during acute exposures

Clinical sign(s)	Escape-impairing?	Humans	Animals	Comments
Piloerection (contraction of erectile tissue of hair follicles, resulting in rough fur)	No		X	
Hypo- or hyperthermia	Yes	X	X	If severe
Excessive perspiration/sweating/panting	No	X	X	Common effects of nerve agents that accompany other effects considered to be escape impairing
Salivation	No	X	X	Common effects of nerve agents that accompany other effects considered to be escape impairing
Syncope (loss of consciousness)	Yes	X	X	If observed
Blurred vision	Yes	X		If severe
Mydriasis (reflex pupillary dilation)	Yes	X	X	If severe
Miosis (constriction of pupil, regardless of light)	Yes	X	X	If severe
Ptosis (drooping of upper eyelids)	Yes	X	X	
Chromodacryorrhea (red lacrimation)	No		X	
Loss of libido	No		X	
Muscle tone/equilibrium				
Abnormal gait or postural observations	Yes	X	X	If sufficient to impair balance or locomotion
Mobility	Yes	X	X	If severely impaired
Righting	Yes		X	If moderately or severely impaired
Forelimb grip strength	Yes		X	If severely impaired
Landing foot splay	Yes		X	If significantly increased a measure of postural instability

(Continued)

Table C–1 (Continued). Common clinical signs, symptoms, and histopathological abnormalities observed during acute exposures

Clinical sign(s)	Escape-impairing?	Humans	Animals	Comments
Fasciculation (muscular twitching)	Yes	X		If severe
Muscle weakness of extremities (foot drop)	Yes	X		If severe
Decreased manual dexterity	Yes	X		If severe
Decreased nerve conduction velocity	Yes	X		If accompanied by changes that affect locomotion
Sensorimotor reactivity				
Click response	No		X	Sensitive response
Touch response	No		X	Sensitive response
Tail pinch response	Yes		X	If no or limited reaction, indicates decreased sensory ability and CNS impairment
Paresthesia (numbness/tingling body parts)	Yes	X		If impairs locomotion or ability to grasp
Perception speed	No	X	X	Sensitive response
Reaction time (simple or choice)	No	X	X	Sensitive response
Auditory vigilance	No	X	X	Sensitive response
Visual time discrimination	No	X	X	Sensitive response
Depth and form perception	No	X	X	Sensitive response
Tinnitus (ear ringing)	No	X		
Pressure in the ears	No	X		
Reduced hearing acuity	No	X		
Insomnia or wake frequently	No	X	X	

(Continued)

Table C–1 (Continued). Common clinical signs, symptoms, and histopathological abnormalities observed during acute exposures

Clinical sign(s)	Escape-impairing?	Humans	Animals	Comments
Nervous system histopathology				
Central nervous system lesions	Yes		X	If degenerative change observed
Peripheral nervous system lesions	Yes		X	If degenerative change observed
Other				
GI tract signs and symptoms				
Stomach ache	Yes	X		If severe—e.g., causes involuntary doubling over
Nausea	No	X		May be accompanied by weakness or dizziness that will be considered escape-impairing
Diarrhea	No	X		
Vomiting	Yes	X	X	If severe
Cardiovascular changes				
Change in blood pressure	Yes	X	X	If severe, assumes may induce faintness or dizziness (extreme hypotension)
Changes in heart rate (tachycardia or bradycardia)	Yes	X	X	If severe or accompanied by other impairing cardiovascular change
Tightness in the chest	Yes	X		If severe or accompanied by other impairing cardiovascular change
Pains in heart or chest	Yes	X		If severe or accompanied by other impairing cardiovascular change
Arrhythmias	Yes	X	X	Assumes sufficient to impair systemic blood flow
Ventricular fibrillation	Yes	X		If observed

Appendix C: References

Abadin H, Murray H, Wheeler J [1998]. The use of hematological effects in the development of minimal risk levels. Reg Toxicol Pharm 28:61–66.

Abadin H, Chou C, Llados F [2007]. Health effects classification and its role in the derivation of minimal risk levels: Immunological effects. Regul Toxicol Pharmacol 47:249–256.

Chou C, Pohl H [2005] Health effects classification and its role in the derivation of minimal risk levels: Renal effects. Regul Toxicol Pharmacol 42:202–208.

DeRosa CT, Stara JF, Durkin PR [1985]. Ranking of chemicals based upon chronic toxicity data. Toxicol Ind Health 1(4):177–192.

Pellegrino R, Viegi G, Brusasco V, Crapo RO, Burgos F, Casaburi R, Coates A, van der Grinten CPM, Gustafsson P, Hankinson J, Jensen R, Johnson DC, MacIntyre N, McKay R, Miller MR, Navajas D, Pedersen OR, Wanger J [2005]. Interpretative strategies for lung function tests. ATS/ERS task force: Standardisation of lung function testing. Eur Respir J American Thoracic Society 26:948–968.

Pohl H, Abadin H [1995]. Utilizing uncertainty factors in minimal risk levels derivation. Reg Toxicol Pharm 22:180–188.

Pohl H, Chou C [2005]. Health effects classification and its role in the derivation of minimal risk levels: Hepatic effects. Regul Toxicol Pharmacol 42:161–171.

Pohl H, Luukinen B, Holler J [2005]. Health effects classification and its role in the derivation of minimal risk levels: Reproductive and endocrine effects. Regul Toxicol Pharmacol 42:209–217.

US EPA (United States Environmental Protection Agency) [1994]. Methods for derivation of inhalation reference concentrations and application of inhalation dosimetry. EPA/600/8-90/066F.

This page intentionally left blank.

APPENDIX D: Analyses Supporting the Development of Uncertainty Factor Approach

During the 1994 update to the IDLH values, NIOSH chose to use an approach for extrapolation from a duration-adjusted LC value (e.g., LC_{50} value) that included the use of a default UF (or safety factor) of 10. When data on effects other than lethality were available, an unspecified UF less than 10 was typically applied. The rationale for this decision was neither described in the 1994 update methodology [NIOSH 1994] nor specified in the support documentation for the IDLH values derived from such data. This has resulted in the need to re-examine this practice to verify its efficacy in the derivation of health-protective IDLH values.

To evaluate the assignment of UFs, NIOSH conducted several analyses in preparation of the IDLH methodology to determine if the approach used in the 1994 update needed to be revised. Numerous datasets were evaluated to identify the typical ratio between the IDLH and the POD derived from different types of studies. The results of these analyses are presented in Section D.1. In arriving at the final methodology presented in this CIB, the empirical analysis discussed in this appendix was supplemented by previously published data analysis [Fowles et al. 1999; Rusch et al. 2010]. Current risk assessment principles related to the rationale and concepts for UF application for setting exposure guidelines of different types were also considered, in particular the process used by the AEGL committee.

D.1 Analysis for Selected Approach

To derive a scientifically based approach for the use of UFs in the derivation of IDLH values, several analyses were conducted to determine the appropriate size of the UF for extrapolating from various points of departure, taking into account the weight-of-evidence approach and MOA considerations described previously. Two approaches were used. Approach 1 involved a detailed evaluation of acute toxicity data for a selection of 20 chemicals, whereas Approach 2 evaluated the MOAs identified from a larger dataset of 94 chemicals.

For Approach 1, 20 case-study compounds with high-quality animal lethality studies and adequate human effects data to estimate lethality thresholds were identified. The Log-Probit model of USEPA's BMDS was used to calculate the LC_{50} and LC_{01} values based on the mortality incidence data for each of the animal studies of adequate quality. All of the animal LC_{50} values and human lethality threshold data were adjusted to 30-minute-equivalent values via the method of ten Berge and colleagues [ten Berge et al. 1986], by using chemical-specific values of n for lethality whenever possible or standard defaults (i.e., $n = 1$ for extrapolation from shorter to longer durations and $n = 3$ for extrapolation from longer to shorter durations), and using an n of 1 for time correction of human effects other than lethality (e.g., irritation or signs of CNS depression). It should be noted that the default ten Berge adjustment approach would also be most appropriate for less than lethal effects. However, since this analysis was intended as one of several range-finding approaches for uncertainty selection explored by NIOSH, the additional analysis required to make the adjustments from the poorly documented human studies was deemed an unnecessary refinement for this particular analysis. The default ten Berge approach is specified for lethal and non-lethal effects within the IDLH methodology outlined in this CIB.

The correct approach for extrapolation is uncertain for less-than-lethal effects. Adequate quantitative data are rarely available for severe adverse effects in humans to support concentration–response modeling. In particular, thresholds for lethality are difficult to estimate from the very limited available case report information. However, available effect levels in humans gleaned from peer-reviewed secondary sources were arrayed by concentration (Conc), duration of exposure (time, t), the concentration × duration product (Conc × t = k), and severity of effect for each study that provided human response data.

Results of this analysis are shown in Table D–1. The analysis found that animal lethal concentrations and human effect thresholds were generally correlated for this limited dataset. Additional analyses were conducted by MOA category (e.g., irritant, CNS depressant, or "other"). Group means for each MOA category were not significantly different when comparing animal lethal concentrations (LC_{50} and LC_{01} values) to human lethality thresholds (human LC_{LO} values). However, group means for the three MOA categories did differ significantly for the ratios of animal lethal concentrations (LC_{50} and LC_{01} values) versus the human LOELs for the 20 case-study chemicals. The mean LC_{50}/human LOEL ratio was greatest for irritants, followed by chemicals that induce CNS effects, and then chemicals that had other MOAs.

As shown below in Table D–1, comparison of animal RD_{50} values to IDLH values suggests that, on average, the RD_{50} corresponds to a human severe irritation threshold, since the IDLH values used in the analysis were based on irritant effects in humans. This interpretation is consistent with study findings [Schaper 1993] that suggested that exposure at the RD_{50} would likely cause intolerable sensory irritation. However, it is noteworthy that the RD_{50} would have been considered in the overall weight of evidence in setting the IDLH values used in our analysis, which might have biased the results toward a value of 1.

Table D–1. Ratio of lethal concentrations from animal studies and observed or estimated human effect levels

Comparison	Median	95th Percentile
LC_{50}/LOEL*	25	330
LC_{01}/LOEL	15	130
LC_{50}/LC_{LO}	2	13
LC_{01}/LC_{LO}	1.5	11
LC_{50}/IDLH value†	8	67
RD_{50}/IDLH value†	1	9

Abbreviation: IDLH = immediately dangerous to life or health; LC_{01} = the statistically derived air concentration that caused lethality in 1% of test animals; LC_{50} = median lethal concentration; LC_{LO} = lowest lethal concentration of a substance in the air reported to cause death; LOEL = lowest observed effect level; RD = respiratory depression.
*Based on analysis of 20 case study substances. The numerator is the value from animal studies and the denominator is the human effect level or value of the IDLH value.
†Based on analysis of IDLH values.

The second approach used data directly from current IDLH value documentation to analyze all of the chemicals in the current list of IDLH values that are based on human effects data and had at least one reported LC_{50} value, resulting in a list of 94 chemicals for further examination. For each of these chemicals, the analysis identified the value of the lowest adequate 30-minute adjusted LC_{50} value, the current IDLH value, and the MOA for which the current IDLH value was set. As for the first approach, three MOA categories were used:

1. Irritation
2. Neurological effects
3. "Other"

It was noted that the "other" category included several pesticides that act via inhibition of cholinesterase. Although this group was not analyzed separately, it does form a potential fourth group for additional analysis. The cholinesterase inhibitors were not included in the general neurological effects category, since they have a specific underlying mechanism that might yield significant differences in lethality to non-lethal-effect ratios, as compared with other organics that act via the more general mechanisms of CNS depression. Published data were also used to compile RD_{50} estimates (the concentration of the chemical that results in a 50% decrease in respiratory rate in a standardized rodent test) for these same chemicals.

The distribution of the LC_{50}/IDLH value ratios is shown in Figure D–1. Results of the LC_{50}/IDLH value ratio analysis (shown in Figure D–1) indicate that a factor of 10 would account for human effect thresholds for effects such as severe irritation and neurological effects, for approximately half of the chemicals reviewed, although a factor as high as 100 may be needed to cover 95% of chemicals. Distribution of RD_{50}/IDLH value ratios for 26 chemicals yielded a median ratio of 1, suggesting that exposure at the RD_{50} would generally result in sensory irritation of sufficient severity to be judged as escape-impairing. This interpretation is consistent with study results [Schaper 1993] suggesting that exposure at the RD_{50} would likely cause intolerable sensory irritation. Overall, no clear pattern regarding MOA was evident when comparing LC_{50}/IDLH value ratios and its primary MOA for the 94 chemicals or comparing RD_{50}/IDLH value ratios for the 26 chemicals.

This analysis hypothesized that potent irritants may have a greater difference between the LC_{50} and the threshold for serious effects in humans, as compared with chemicals that cause toxicity via other modes of action. If this hypothesis was true, then the implication would be that deriving an IDLH value from an LC_{50} for such chemicals would require a greater UF than would be needed for chemicals with other modes of action. The analysis produced mixed results, with a significant MOA effect observed for a subset of 20 chemicals, but not in a broader analysis of current IDLH values. Based on these results, the data are not adequate to recommend a different UF by MOA category.

D.2 Recommendation for Deriving IDLH Values

Three primary methods are traditionally applied during the development of acute emergency limits, such as the IDLH values, ERPGs, and AEGLs, to account for uncertainty in extrapolating from a key study to arrive at the final value. In developing the IDLH methodology, three possible approaches were considered.

- **Method 1—Use a weight-of-evidence approach, without specifying any default UF values.** This would be an approach consistent with many volunteer groups that set acute occupational values (e.g., the AIHA ERPG committee). This approach provides for the greatest degree of flexibility in integrating all the complexities of the data, without having to explain departures from defaults that might not be very meaningful in the context of a specific dataset. However, this approach generally has limitations in that it is not highly transparent—i.e., it is often difficult to "back-calculate" the basis for the final numeric value.

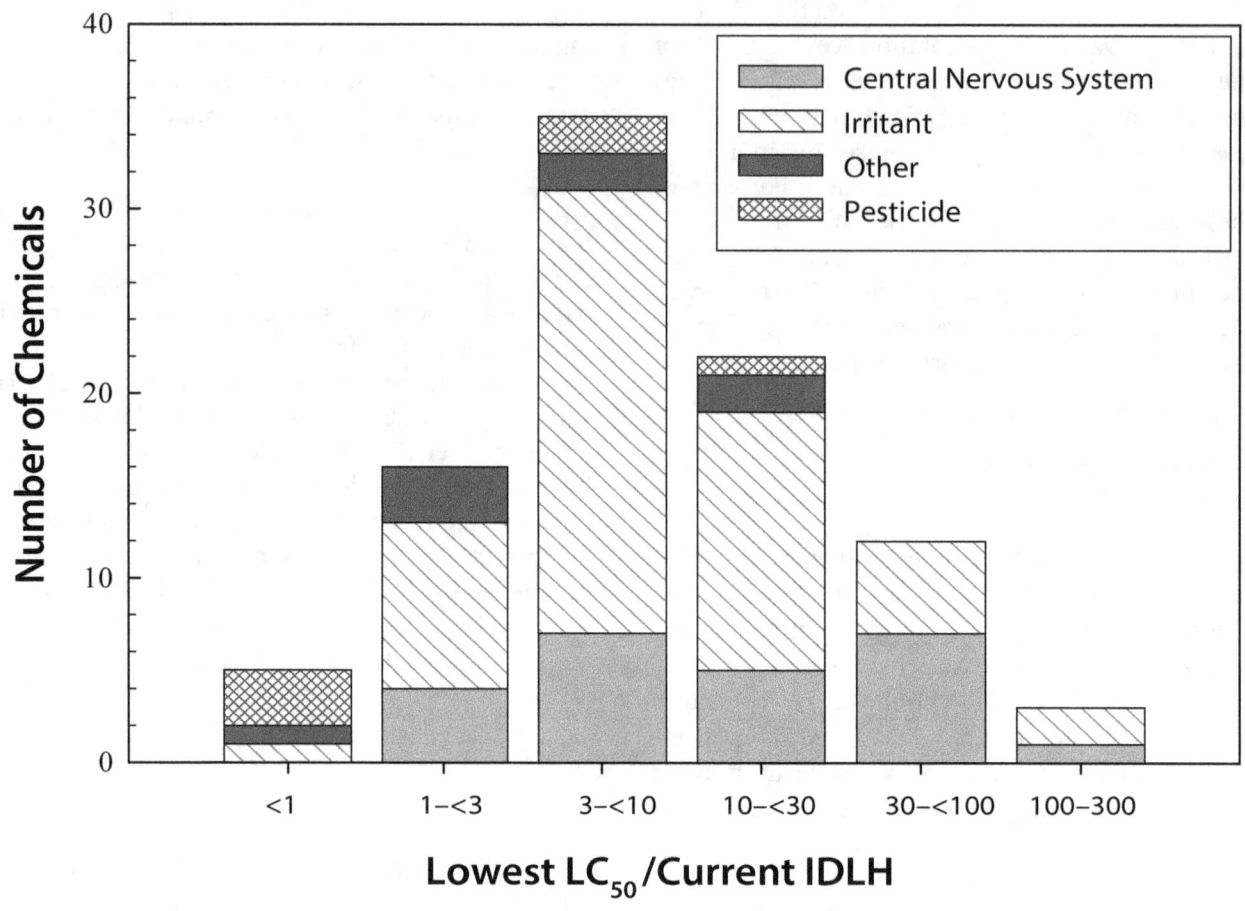

LC_{50}—Concentration to cause a 50% mortality rate in an acute toxicity study.
Irritants—The critical effect that would be the basis for an IDLH value is irritation.
CNS Depressants—The critical effect that would be the basis for an IDLH value is CNS system depression.
Other—The critical effect that would be the basis for an IDLH value arises from an MOA other than irritation or CNS depression.
Pesticide—The critical effect that would be the basis for an IDLH value is cholinesterase inhibition.

Figure D–1. The distribution of ratios of the lowest 30-minute adjusted LC_{50} value to the current IDLH value is shown for 94 substances, representing four MOA categories, to evaluate the potential uncertainty value that provides adequate coverage for each MOA.

- **Method 2—Use a preliminary composite UF as a starting point, based on the nature of the overall dataset, and communicate areas of uncertainty that impacted the final IDLH value in the rationale statement.** This approach provides a data-informed starting point for the analysis, supported by empirical analysis (see Section D.2). The approach is intended to provide flexibility in UF selection by accounting for typical overlaps in individual UF and data hierarchies at the beginning of the UF selection process. This provides an increase in transparency over the weight-of-evidence approach, without requiring significant effort to explain departures from prescribed defaults. This is the approach that has been included in the IDLH methodology.

- **Method 3—Apply a set of default UFs and revise post-hoc on the basis of the dataset.** This would be an approach similar to that used to derive the AEGL values. This approach assigns default values for well-defined areas of uncertainty that pertain to a specific dataset. The final UF is derived by multiplying the individual factors. This approach is the most clear in terms of transparency (i.e., ability to back-calculate the derived value). However, because of the nature of the datasets involved, application of default values often yields conflicting or inappropriate values from one potential critical study to another. The end result of such data conflicts is the application of a post-hoc weight-of-evidence evaluation, in which the final UF or critical study selected might be changed to align better with the overall dataset.

Application of the three methods outlined above should yield similar results, with the primary differences focusing on the level of transparency offered by each approach versus the need for post-hoc modifications. There has been a history of successful application of methods 1 and 3, in the context of acute emergency limit-setting. Method 2 is a hybrid of the methods that attempts to incorporate previous data and experience tailored to the unique needs of the IDLH Program. Method 2 is recommended as a reasonable blend of providing transparency in the basis for an assessment, without the rigid application of default values that may require extensive post-hoc explanations. Multiplication of default UF values may also tend to yield IDLH values that are more than adequately protective. In developing the approach, it was considered that setting IDLH values lower than needed can present new safety risks in the context of the intended application as a tool for respiratory protection selection.

Appendix D: References

Fowles JR, Alexeeff GV, Dodge D [1999]. The use of benchmark dose methodology with acute inhalation lethality data. Regul Toxicol Pharmacol 29(3):262–278.

NIOSH [1994]. Documentation for immediately dangerous to life or health concentrations (IDLH). Cincinnati, OH: U.S. Department of Health and Human Services, Centers for Disease Control and Prevention, National Institute for Occupational Safety and Health. NTIS Publication No. PB-94-195047.

Rusch GM, Bast CB, Cavender FL [2009]. Establishing a point of departure for risk assessment using acute inhalation toxicity data. Regul Toxicol Pharmacol 54(3):247–255.

Schaper M [1993]. Development of a database for sensory irritants and its use in establishing occupational exposure limits. Am Ind Hyg Assoc J 54(9):488–544.

ten Berge WF, Zwart A, Appelman LM [1986]. Concentration-time mortality response relationship of irritant and systematically acting vapors and gases. J Haz Mat 13:301–309.

This page intentionally left blank.

APPENDIX E: Quantitative Adjustments during the Derivation of IDLH Values

This appendix provides supplemental information on quantitative adjustments applied within the IDLH methodology. Section E.1 discusses considerations applied when route-to-route extrapolation is conducted during the derivation of an IDLH value. Section E.2 provides supplemental detail pertaining to adjustments made during the time scaling of data.

E.1 Inhalation Volume Adjustments Approach for Route-to-Route Extrapolation

During the 1994 update of the IDLH values, a volume of 10 m³ for inhaled air was included in the methodology as the basis for calculating the inhalation equivalent concentration from an oral toxicity study dose. This value represents the volume of air assumed to be inhaled over the course of a typical 8-hour shift at light work load. The IDLH methodology uses an alternative approach that takes into account that IDLH values are intended to be based on a maximum exposure duration of 30 minutes. Assuming a worker breathing rate of 50 L/min for 30 minutes, the corresponding total inhaled volume of air for a scenario relevant to the IDLH would be 1.5 m³. This volume corresponds with dosimetry guidelines recommended by the International Commission on Radiological Protection [ICRP 1994]. Use of the value 1.5 m³ in the calculation of the inhalation equivalent concentration from an oral toxicity study dose has been applied in the IDLH method as the default for route-to-route calculations, as discussed in Section 3.4.2.4.

The inhalation concentration calculated with use of the 1994 methodology is built on more health-protective assumptions (yields a lower IDLH value) than the IDLH methodology, and it may have resulted in IDLH values that are more than adequately protective in some cases. The impact of this change in the assumed inhaled volume can be highlighted in the context of a typical example for a chemical with an IDLH value based on systemic toxicity derived from a single-oral-dose acute lethality study. Table E–1 illustrates the differences between the 1994 approach that used 10 m³ and the IDLH methodology presented in the CIB that uses 1.5 m³.

Because the impact of the selected default is significant (roughly 7-fold), additional insight into the conditions under which each possibility is most scientifically appropriate is needed. Factors that impact this judgment include these:

- The toxicokinetic profile associated with the dosing regimen for the critical study compared to the IDLH acute inhalation scenario. Key considerations include 1) differences in absorption kinetics and 2) differences in clearance or elimination kinetics.

- The MOA and its associated dose metric for the adverse effect, commonly based on either peak (maximum) concentration (Cmax) or total dose (represented by the area under the curve [AUC]).

Under the most common scenarios and datasets available for setting IDLH values, the 1.5 m³ assumption is likely to be adequate. The level of concern about the degree to which the 1.5 m³ value is likely to be protective can be summarized for different scenarios. If the MOA is based on the peak concentration, then the differences in absorption and

Table E–1. Illustration of the impact of various inhaled air volumes on derived IDLH values

Approach	Oral LD$_{50}$ value (mg/kg)	Body weight (kg)	Inhaled air volume (m³)	UF*	Derived IDLH value[†] (mg/m³)
1994 Approach	50	70	10	100	3.5
IDLH Methodology	50	70	1.5	100	23.3

Abbreviation: IDLH = immediately dangerous to life or health; LD$_{50}$ = median lethal dose; mg/kg = milligrams per kilograms of body weight; m³ = cubic meters; UF = uncertainty factor

*The UF of 100 is included as an example only and is not a default value used during the derivation of IDLH values based on oral data.

[†]Derived IDLH values were calculated with use of the following equation: IDLH value = [(Oral LD$_{50}$ value * Body weight)/Inhaled Air Volume)]/(UF)

elimination rate kinetics for the dosing regimen used in the non-inhalation study can impact the selection of an inhalation volume. Simplifying assumptions related to the absorption rates of alternative dosing routes and methods can be applied to generalize the decision process:

- Low Concern:
  - If the underlying MOA for the critical effect is based on total systemic dose (AUC) and dose is sufficiently low that absorption and elimination rates are not severely saturated. In this case the total systemic dose (e.g., total mg dose) is the driver for the response, not the rate of uptake, and thus the most appropriate calculation of an IDLH yields an air concentration associated with achieving the total mass dose (mg) over the 30-min IDLH duration.
  - If the underlying MOA is based on peak concentration, an IDLH based on oral gavage dosing regimen should ensure that the chemical-specific kinetics allow for rapid absorption in the GI tract (short Tmax). In this case, the peak concentration achieved from a 30-min exposure by either route is approximately equal.
  - AUC or peak concentration is correct dose metric and the critical acute study is based on IV or IP injection data. Uptake is considered immediate and would parallel assumed rapid uptake via inhalation; thus, peak doses would be roughly similar across routes.

- Moderate Concern:
  - AUC is the correct dose metric and an oral dosing study was used. In most cases, since total dose is the driver for the onset of toxicity, use of the acute inhalation volume will be appropriate. An exception is the absorption or elimination rate is severely saturated, such that the orally administered dose does not represent well the total dose that would be received following an acute inhalation exposure at the IDLH. Under such conditions, the actual internal dose received by the test animals is lower than estimated by the administered oral dose, and the percentage of systemic bioavailability of the toxic form of the chemical (parent or metabolite) would need to be considered. Direct adjustment of the animal dose would be the best approach under these conditions. However, using some greater inhalation intake time to account for higher systemic doses across the two routes could also be considered in the absence of quantitative kinetic adjustments based on equivalent systemic doses across routes.
  - Peak concentration is the correct dose metric and the chemical has slow

absorption kinetics in an oral gavage dosing study, or the critical oral study used serial dosing or continuous dosing protocols. In these cases the peak concentration in the critical study would not represent the likely peak concentration reached for the inhalation study, and the currently proposed extrapolation method would not necessarily be adequately protective.

The toxicokinetic considerations for route-to-route extrapolation are complex. In most cases, because of the nature of the acute systemic effects involved in IDLH derivation, rapid absorption kinetics and rapid onset of effects are expected. Thus, under the most likely conditions, the 30-min inhalation volume of 1.5 m^3 is viewed as an adequate default approach. However, there are scenarios based on chemical kinetics or non-inhalation study designs that may impact the level of protection afforded by the default adjustment. For this reason, the IDLH methodology has been modified to further communicate the potential conditions where additional kinetic-based adjustments may be needed.

E.2 Time Scaling Adjustments

In most cases IDLH values are derived from studies having exposures for periods shorter than or longer than 30 minutes. Thus, the PODs derived from such studies are adjusted to 30-minute-equivalent values. This adjustment is made by using the ten Berge et al. [1986] modification to Haber's Rule that assumes the following relationship between concentration (Conc) and duration (time, t): (Conc^n × t = k). The impact of the value of n on the shape of the concentration–time–response curve is shown in Figures E–1 and E–2. As shown in these figures, larger values of n result in flatter curves, meaning that, for a given degree of toxicity, the concentration varies less with changes in duration. This is particularly apparent in Figure E–1, which shows the extrapolation from 4 hours to 30 minutes. This figure shows the impact of using different values of n to extrapolate to shorter durations from a concentration of 10 ppm at 4 hours. In this example,

an n of 3 results in a concentration at 30 minutes that is not much higher than the test concentration at 4 hours, whereas the calculated concentration at 30 minutes is substantially higher when n = 1. Thus, using n = 3 for extrapolating from longer durations to 30 minutes results in lower concentrations, a more health-protective approach.

Figure E–2 shows the converse situation, extrapolating from an exposure to 10 ppm for 15 minutes to longer durations. In this case, the steeper curve associated with n = 1 results in a lower concentration at 30 minutes, compared with the value calculated using n = 3. Thus, using n = 1 is a more health-protective approach in extrapolating from shorter durations to 30 minutes.

Based on these considerations, a default value of n = 1 is used for extrapolation from shorter durations, and a default value of n = 3 is used for extrapolation from longer durations to the 30-minute duration of interest. In both cases, a calculated n specific to the chemical and species of interest is preferred when data are available to calculate the value.

The data used to construct Figure E–1 are shown in Table E–2. Table E–2 shows the calculated concentrations when extrapolating from 10 ppm at 4 hours, using n values of 1, 2, or 3.

The data used to construct Figure E–2 are shown in Table E–3. Table E–3 shows the calculated concentrations when extrapolating from 10 ppm at 0.25 hours, using n values of 1, 2, or 3.

The following paragraph illustrates the effects of time scaling on inhalation toxicity data evaluated during the development of IDLH values for three chemicals:

- 1,1-Dimethylhydrazine (CAS# 57-41-7)
- Vinyl acetate (CAS# 108-05-04)
- Titanium tetrachloride (CAS# 7550-45-0).

In the first example, the identified LC$_{50}$ and LOAEL values for 1,1-dimethylhydrazine correlated to exposure durations of 5 or 15 minutes. No empirically derived n values were identified within the

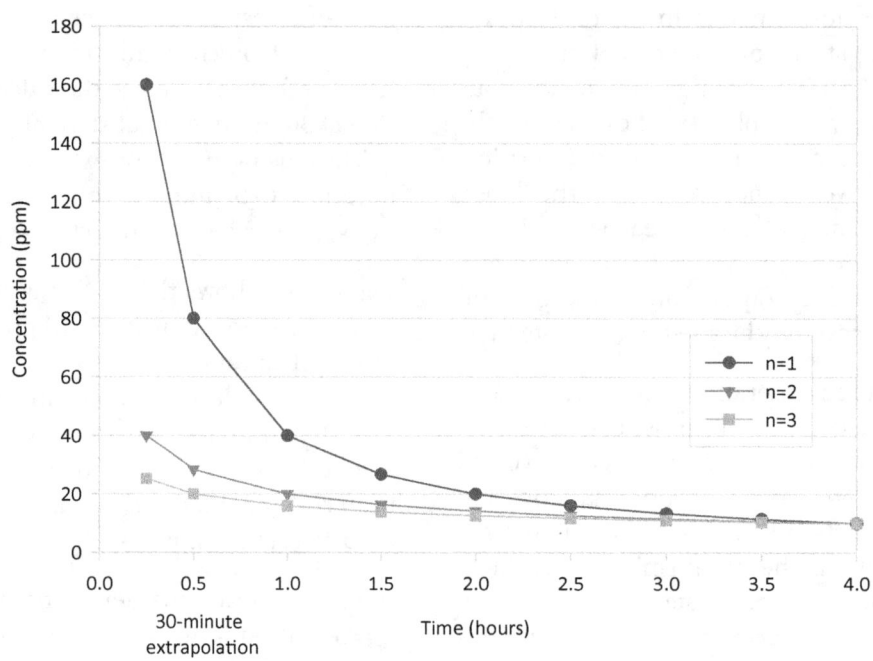

*n = exponent applied within ten Berge equation [1986]

Figure E–1. ten Berge Extrapolation from longer (4 hour) to shorter (30 minute) durations

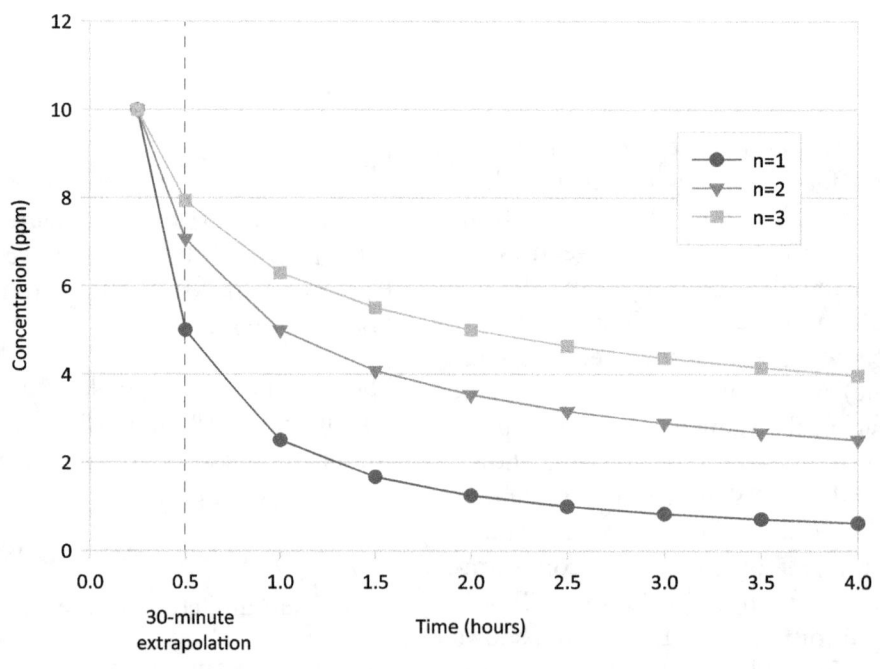

*n = exponent applied within ten Berge equation [1986]

Figure E–2. ten Berge extrapolation from shorter (15 minute) to longer (30 minute) durations

Table E–2. Time scaling for 10 ppm at 4 hours

Time, hours	n* = 1	n = 2	n = 3
0.25	160	40	25
0.5	80	28	20
1	40	20	16
1.5	27	16	14
2	20	14	13
2.5	16	13	12
3	13	12	11
3.5	11	11	10
4	10	10	10

n* = exponent applied within ten Berge equation [1986]

Table E–3. Time scaling for 10 ppm at 15 minutes (0.25 hours)

Time	n* = 1	n = 2	n = 3
0.25	10	10	10
0.5	5	7	8
1	3	5	6
1.5	2	4	6
2	1	4	5
2.5	1	3	5
3	1	3	4
3.5	1	3	4
4	1	3	4

n* = exponent applied within ten Berge equation [1986]

reviewed literature for 1,1-dimethylhydrazine. Because the selected data were associated with exposure times less than 30 minutes, a default value of 1 for n within the ten Berge equation was applied on the basis of the rationale discussed in the previous paragraphs to extrapolate the most health-protective estimate. Time scaling resulted in a reduction of the exposure concentrations to approximately 17% to 50% of the original exposure concentrations for the 5- and 15-minute durations, respectively. Table E–4 provides the extrapolated 30-minute-equivalent concentration for 1,1-dimethylhydrazine. Figure E–3 provides a visual representation of the data for 1,1-dimethylhydrazine. In comparison, the selected LC_{50} and LOAEL values for vinyl acetate were associated with exposure durations of 2 to 6 hours. Because no empirically derived value of n was available, a default value of 3 for n was used for time scaling in the ten Berge equation to adjust the data points from longer to shorter exposure durations. As noted earlier, this is a health-protective default. The resulting extrapolated concentrations were approximately double the original exposure concentrations and can be found in Table E–5 and Figure E–4. The last example, titanium tetrachloride (see Table E–6 and Figure E–5), demonstrates the effects of the use of an empirically derived n to calculate the 30-minute equivalents for exposure concentrations associated with durations both shorter and longer than 30 minutes. A value of 0.88 has previously been calculated by the NAS during the development of the AEGL values for titanium tetrachloride [NAS 2007]. For data corresponding to exposure durations less than 30 minutes, the resulting extrapolated concentrations were approximately 5% to 50% of the original LC_{50} and LOAEL values. Substantial changes in the extrapolated 30-minute-equivalent concentrations were also observed when extrapolating from longer to shorter durations, with the relative increases being in a range of 2 to 10 times higher than the original value. As evident by the three previous examples, selection of the appropriate n during time scaling may greatly affect the resulting 30-minute-equivalent concentrations.

Table E–4. Time scaling of toxicity data from shorter to longer durations, using the ten Berge equation: 1,1-dimethylhydrazine (CAS# 57-41-7)

Study no.	Species	Toxicological endpoint LC$_{50}$ value (ppm)	LOAEL (ppm)	Exposure duration (minute)	n	30-minute equivalent value (ppm)
1	Rat	8,230		15	1	4,115
2	Rat	24,500		5	1	4,083
3	Dog	3,580		15	1	1,790
4	Dog	22,300		5	1	3,717
5	Dog		360	15	1	180
6	Dog		1,550	5	1	258

Abbreviations: LC$_{50}$ value = median lethal concentration; LOAEL = lowest observed adverse effect level; ppm = parts per million; n = exponent applied in ten Berge equation [1986]. The study data are from Weeks et al. [1963].

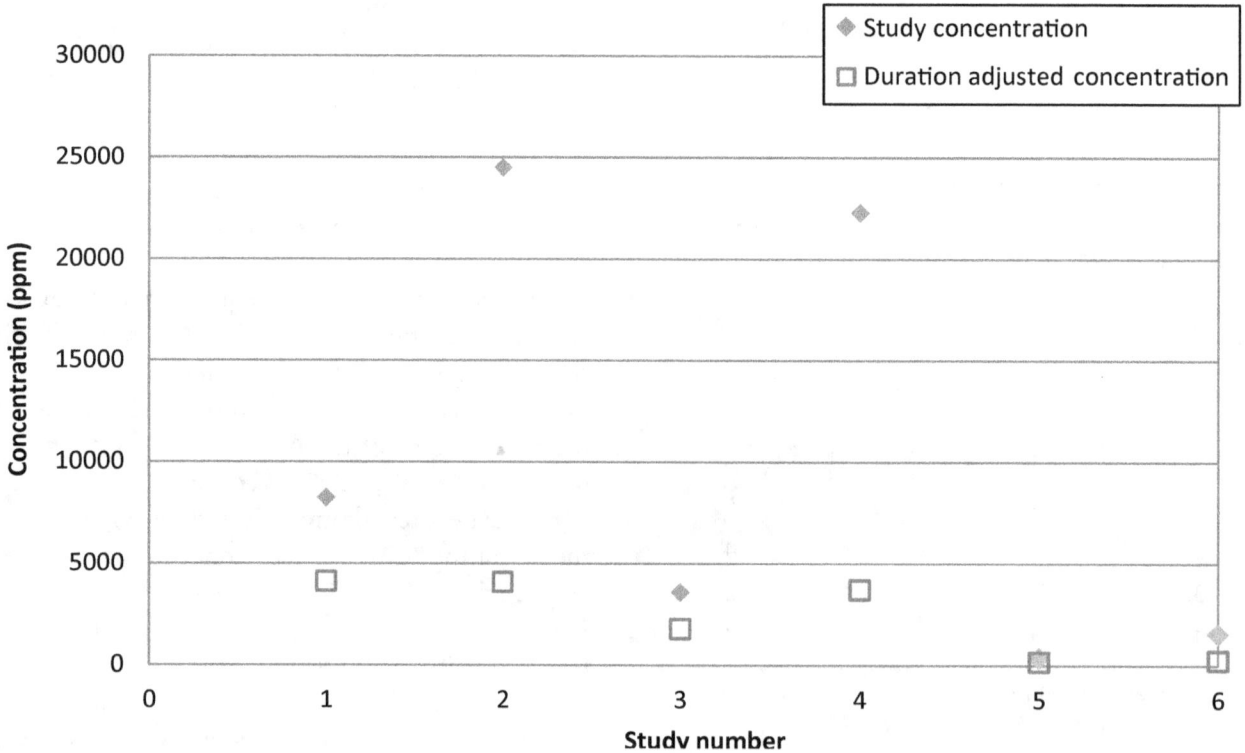

Figure E–3. Time scaling of toxicity data from shorter to longer durations, using the ten Berge equation: 1,1-dimethylhydrazine (CAS# 57-41-7)*

Abbreviation: ppm = parts per million
*All 30-minute data points are duration-adjusted values.

Table E–5. Time scaling of toxicity data from shorter to longer durations, using the ten Berge equation: vinyl acetate (CAS# 108-05-04)

Study no.	Species	Reference	Toxicological endpoint		Exposure duration (minute)	n	30-minute-equivalent value (ppm)
			LC_{50} (ppm)	LOAEL (ppm)			
1	Rat	Bogdanffy et al. [1997]		1,000	360	3	2,289
2	Rat	Roumiantsev et al. [1981]	3,238		240	3	6,476
3	Mouse	Smyth and Carpenter [1973]	1,460		240	3	2,920
4	Dog	Smyth and Carpenter [1973]		3,280	240	3	6,274
5	Guinea pig	Smyth and Carpenter [1973]	5,210		240	3	10,420

Abbreviations: LC_{50} value = median lethal concentration; LOAEL = lowest observed adverse effect level; n = exponent applied in ten Berge equation [1986]; ppm = parts per million.

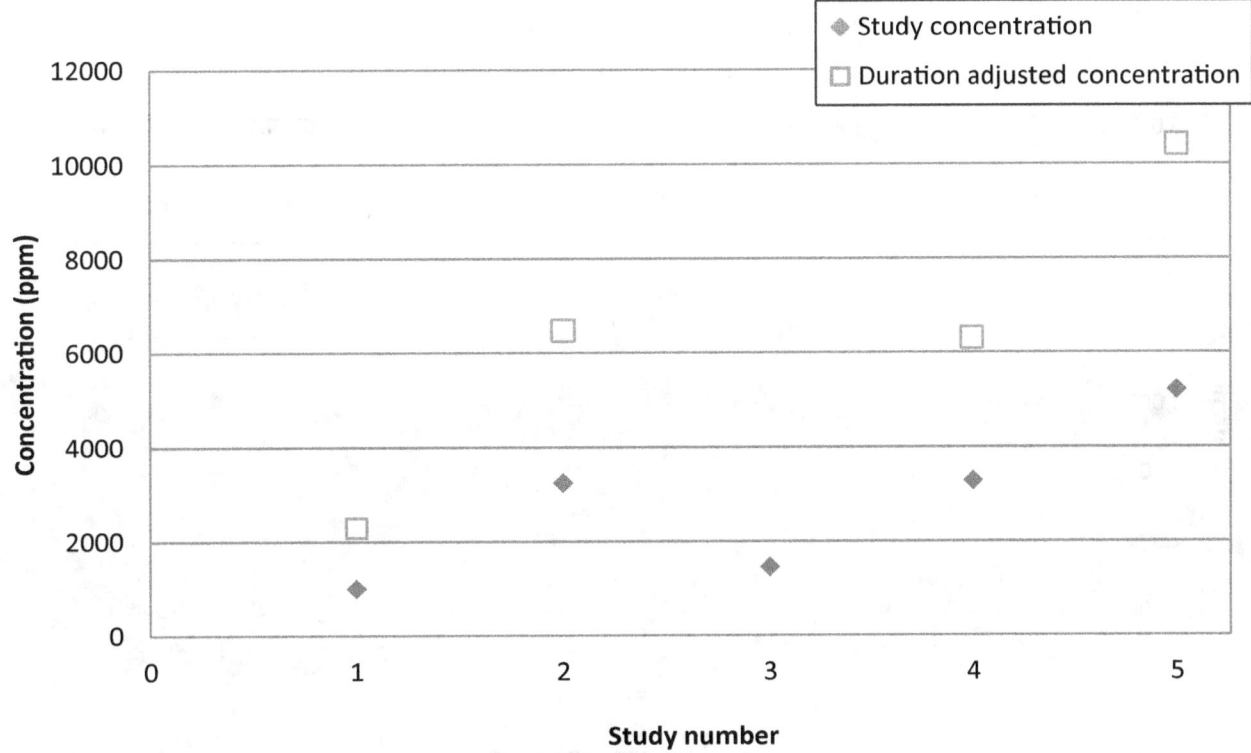

Figure E–4. Time scaling of toxicity data from shorter to longer durations, using the ten Berge equation: vinyl acetate (CAS# 108-05-04)*

Abbreviation: ppm = parts per million
*All 30-minute data points are duration-adjusted values.

Table E–6. Effects of an empirically-derived n on time adjustments, using the ten Berge equation: titanium tetrachloride (CAS# 7550-45-0)

Study no.	Species	Reference	Toxicological endpoint LC_{50} (ppm)	LOAEL (ppm)	Exposure duration (minute)	n	30-minute-equivalent value (ppm)
1	Rat	Kelly [1980]	13,940		2	0.88	642
2	Rat	Kelly [1980]	4600		5	0.88	600
3	Rat	Kelly [1980]	713		15	0.88	324
4	Rat	Kelly [1980]	171		60	0.88	376
5	Rat	Kelly [1980]	143		120	0.88	691
6	Rat	Kelly [1980]	59		240	0.88	627
7	Rat	Gardner [1980]		26	20	0.88	16

Abbreviations: LC_{50} = statistically determined median concentration of a substance in the air that is estimated to cause death in 50% (one half) of the test animals; LOAEL = lowest observed adverse effect level; n = exponent applied in ten Berge equation [1986; NAS 2007]; ppm= parts per million.

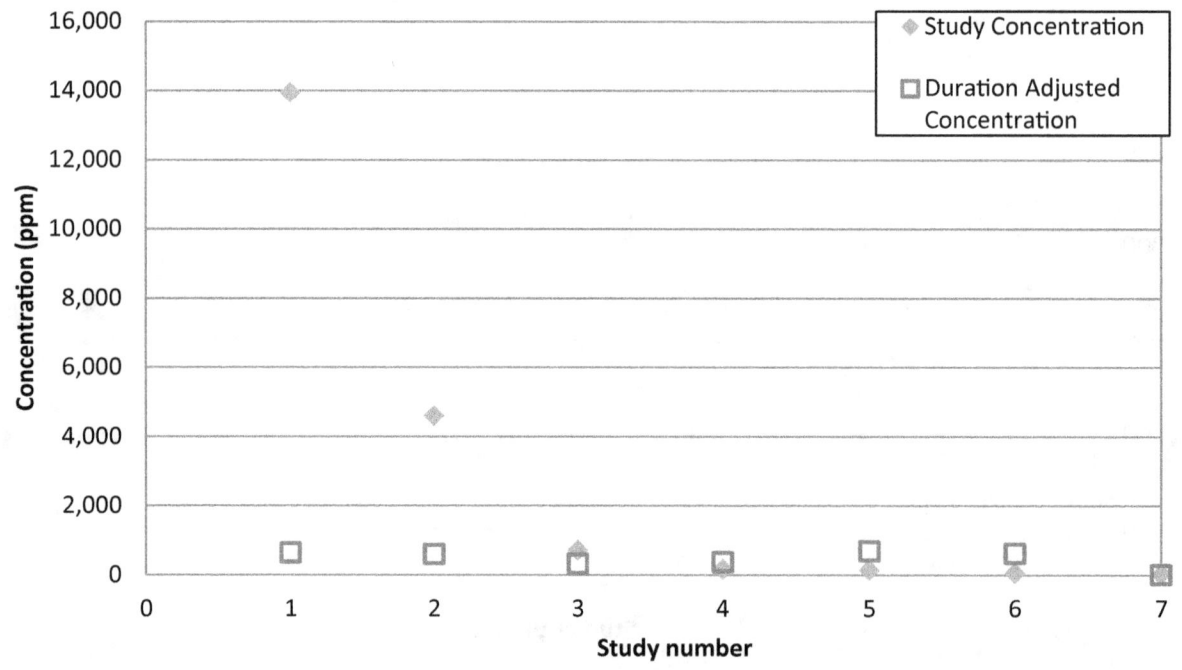

Figure E–5. Effects of an empirically derived n on time adjustments, using the ten Berge equation: titanium tetrachloride (CAS# 7550-45-0)*

Abbreviation: ppm = parts per million.
*All 30-minute data points are duration-adjusted values.

Appendix E: References

Bogdanffy MS, Gladnick NL, Kegelman T, Frame SR [1997]. Four-week inhalation cell proliferation study of the effects of vinyl acetate on rat nasal epithelium. Inhal Toxicol 9:331–350.

Gardner RJ [1980]. Rat sensory irritation study: $TiCl_4$. Haskell report no. 593-80, August 11, 1980. Wilmington, DE: E.I. du Pont de Nemours and Company, Haskell Laboratory for Toxicology and Industrial Medicine.

ICRP [1994]. Human respiratory tract model for radiological protection. ICRP publication 66. Ann ICRP 24(1–3).

Kelly DP [1980]. Acute inhalation studies with titanium tetrachloride. Haskell Laboratory report no. 658–80, October 31, 1980. Wilmington, DE: E.I. du Pont de Nemours and Company, Haskell Laboratory for Toxicology and Industrial Medicine.

NAS [2007]. Interim acute exposure guideline levels (AEGLs) for selected metal phosphides. Committee on Toxicology, Board on Environmental Studies and Toxicology, National Research Council, National Academy of Science. Washington, DC: National Academy Press [http://www.nap.edu/].

Roumiantsev AP, Tiunova LV, Astapova CA, Kustova ZR, Lobanova IA, Ostroumova NA, Petushkov NM, Chernikova VV [1981]. Information from the Soviet Toxicology Center: toxicometric parameters of vinyl acetate. Gig Tr Prof Zabol 11:57–60.

Smyth HF, Carpenter CP [1973]. Initial submission: vinyl acetate: single animal inhalation and human sensory response, with cover letter dated 082792. Doc. no. 88-920010328. Pittsburgh, PA: Carnegie-Mellon Institute.

ten Berge WF, Zwart A, Appelman LM [1986]. Concentration-time mortality response relationship of irritant and systematically acting vapors and gases. J Haz Mat 13:301–309.

Weeks MH, Maxey GC, Sicks ME, Greene EA [1963]. Vapor toxicity of UDMH in rats and dogs from short exposures. Am Ind Hyg Assoc J 24:137–143.

www.ingramcontent.com/pod-product-compliance
Lightning Source LLC
Chambersburg PA
CBHW081730170526

45167CB00009B/3763